T0296189

CAMBRIDGE MONOGRAPHS ON PHYSICS

GENERAL EDITORS

N. FEATHER, F.R.S.
Professor of Natural Philosophy in the University of Edinburgh

D. SHOENBERG, PH.D.
Fellow of Gonville and Caius College, Cambridge

THE AUGER EFFECT AND OTHER RADIATIONLESS TRANSITIONS

THE AUGER EFFECT
AND
OTHER RADIATIONLESS
TRANSITIONS

BY

E. H. S. BURHOP

Reader in Physics, University of London
(University College)

CAMBRIDGE
AT THE UNIVERSITY PRESS
1952

CAMBRIDGE
UNIVERSITY PRESS

University Printing House, Cambridge CB2 8BS, United Kingdom

Published in the United States of America by Cambridge University Press, New York

Cambridge University Press is part of the University of Cambridge.

It furthers the University's mission by disseminating knowledge in the pursuit of education, learning and research at the highest international levels of excellence.

www.cambridge.org
Information on this title: www.cambridge.org/9781107641105

© Cambridge University Press 1952

First published 1952
First paperback edition 2014

A catalogue record for this publication is available from the British Library

ISBN 978-1-107-64110-5 Paperback

GENERAL PREFACE

The Cambridge Physical Tracts, out of which this series of Monographs has developed, were planned and originally published in a period when book production was a fairly rapid process. Unfortunately, that is no longer so, and to meet the new situation a change of title and a slight change of emphasis have been decided on. The major aim of the series will still be the presentation of the results of recent research, but individual volumes will be somewhat more substantial, and more comprehensive in scope, than were the volumes of the older series. This will be true, in many cases, of new editions of the Tracts, as these are republished in the expanded series, and it will be true in most cases of the Monographs which have been written since the War or are still to be written.

The aim will be that the series as a whole shall remain representative of the entire field of pure physics, but it will occasion no surprise if, during the next few years, the subject of nuclear physics claims a large share of attention. Only in this way can justice be done to the enormous advances in this field of research over the War years.

N. F.
D. S.

CONTENTS

CHAPTER IV

The Auger Effect and X-ray Spectra

CHAPTER V

The Internal Conversion of γ-Rays

CHAPTER VI

Internal Conversion Processes in the Creation and Annihilation of Electron Pairs

CHAPTER VII

The Auger Effect in Meson Capture

CHAPTER VIII

Radiationless Transitions in Atomic and Molecular Spectra

EXPLANATION OF PLATES

PLATE I (*facing p. 2*)

Cloud-chamber photographs obtained by Auger (1926) showing forked (multiple) tracks produced by X-ray ionization in neon, argon, krypton and xenon. The exciting X-radiation varied in quantum energy between 20 and 60 keV.

(*a*) Neon ($h\nu_K = 800$ eV.). Single photoelectron tracks only are visible. The tracks show heavier ionization at the origin but no definite Auger tracks are detectable.

(*b*) Argon. Most of the photoelectron tracks are accompanied by short tracks due to Auger electrons of energy about 3000 eV.

(*c*) Krypton. The event in the centre of this plate shows, in addition to the photoelectron, an Auger electron of energy 10 keV. due to internal conversion of K-series radiation in the L shell, and two short tracks of Auger electrons due to internal conversion of L-series radiation in the M shell.

(*d*) Xenon. The interpretation of the track on the left of this photograph is similar to that given under (*c*). The two tracks due to L-series conversion in the M shell are much longer than in krypton, these electrons now having energies of 5 keV.

The notation used to describe the tracks is explained in §3.1. The position and direction of the X-ray beam is indicated by the horizontal arrows. The vertical lines indicate the origins of the electron tracks.

PLATE II (*facing p. 24*)

Cloud-chamber photographs of Auger electrons obtained by L. H. Martin and associates.

(*a*) Stereoscopic pairs of photographs showing Auger electrons and photoelectrons ejected in xenon. Wave-length of exciting X-rays = 0·2Å.

(*b*) Stereoscopic pair showing Auger electrons and photoelectrons from krypton. Wave-length of exciting X-rays = 0·4Å.

The position and direction of the X-ray beam is indicated by the horizontal arrows. The vertical arrows indicate the origins of the electron tracks. The notation used to describe tracks is explained in the text. The angle between the cameras for the stereoscopic pairs was 90°.

PLATE III (*facing p. 78*)

Some typical X-ray L series spectra obtained by Cauchois with a curved crystal spectrometer showing diagram lines, and Coster-Kronig satellites. (*a*) Au. (*b*) Tl. (*c*) Pb.

PLATE IV (*facing p.* 162)

Auger electrons ejected following the capture of μ-mesons by atoms in a photographic emulsion (Cosyns *et al.* 1949).

(*a*) Single slow (Auger) electron ejected at end of meson track.

(*b*) Several slow (Auger) electrons and single fast (decay) electron at end of meson track.

PLATE V (*facing p.* 170)

Auger electrons ejected following the capture of π-mesons by atoms.

In both cases the π-mesons captured by the nucleus produce a star.

In (*a*) there are two and (*b*) a single slow electron at the point of origin of the star. These are interpreted as Auger electrons.

AUTHOR'S PREFACE

In the early years of the present century the study of radiationless transitions following ionization of atoms in inner levels provided important confirmatory evidence for the Bohr-Rutherford atom. Later the quantitative interpretation of transition rates for these processes constituted not the least of the many spectacular achievements of quantum mechanics.

The importance of Auger transitions in the detailed interpretation of X-ray spectra is now generally recognized. Interest in the subject has recently been renewed in connexion with the study of artificial radioactivity and of the possibility of the production of inner shell ionization by the processes of K capture and the internal conversion of γ radiation.

The present work aims at giving a reasonably complete discussion of the Auger Effect. But owing to its close relationship to the Auger Effect it was felt that a discussion of the internal conversion of γ-radiation would also fit naturally into the framework of the present book. Owing to the rapid rate of development it is difficult to be up-to-date in this field, but it is hoped that Chapter v gives a fairly complete picture of the position up to December 1950.

Other topics treated include internal pair production and our present rather meagre knowledge of the rate of Auger processes in meson capture. Finally, Chapter viii discusses radiationless transition processes in atomic and molecular spectra. The discussion of radiationless transitions in molecular spectra is far from complete, as a full discussion of this topic as well as of the many other types of radiationless processes generally classed as collisions of the second kind would have changed completely the balance of the present work.

I am greatly indebted to Professor H. S. W. Massey for his interest and encouragement and to Professor N. Feather who read an early draft and made many valuable suggestions. I would also like to thank Dr D. K. Butt for bringing to my attention recent work on the production of homogeneous positrons published in Russian journals. Original photographs from which plates have been prepared were kindly supplied by Professor L. H. Martin (Plate II), Professor Y. Cauchois (Plate III), Dr C. G. Dilworth (Plate IV) and Dr O. Rochat (Plate V). I am very grateful to them for their assistance in this direction.

Finally I must thank the Cambridge University Press for their helpfulness and the high standard they have maintained on the technical side of the publication.

E. H. S. B.

CHAPTER I

INTRODUCTORY

1.1. The discovery of the Auger effect

When a neutral atom is ionized in an inner shell the system consisting of the ionized atom together with the ejected electron at rest at infinity may be characterized by a positive energy

$$E_S(S = K, L_I, L_{II}, L_{III}, M_I, ...),$$

depending upon the level from which the electron has been ejected. This energy, known as the ionization energy for the level S, is measured relative to the energy of the neutral atom in its ground state. A transition may occur in which the inner vacancy is filled by an electron from a higher (less tightly bound) level S', the excess energy appearing as radiation of frequency ν given by

$$h\nu = E_S - E_{S'}. \tag{1.1}$$

Alternatively, the reorganization may occur without emission of radiation. In this case the energy is communicated to another electron of the same atom and this electron is ejected. If

$$E'_R(R = L_I, L_{II}, L_{III}, M_I, ...)$$

is the ionization energy of this electron, it will leave the atom with kinetic energy, T_A, given by

$$T_A = E_S - E_{S'} - E'_R. \tag{1.2}$$

Since E'_R is the ionization energy appropriate to an atom already singly ionized in an inner shell, $E'_R > E_R$. If Z is the atomic number, E'_R is nearly the ionization energy of the level R of an atom of atomic number $Z+1$, so that in general we may write, approximately, $E'_R(Z) = E_R(Z+1)$.* Clearly the process described by (1.2) will be energetically possible only if

$$E_S - E_{S'} > E'_R.$$

As a result of this condition the level R cannot be a K level.

---

* See, for example, Table VIII, §3.8, and Table XIV, §4.6.

The process of radiationless reorganization of an atom ionized
in an inner shell is usually known as the Auger effect, and the
ejected electrons are known as Auger electrons. This is in recogni-
tion of the work of Auger (1925), who first interpreted, in terms of
this process, the paired tracks obtained in a Wilson expansion
chamber containing inert gases ionized by a beam of X-rays.

Plate I shows some beautiful examples of cloud-chamber pictures obtained by Auger (1925, 1926). The paired tracks have a
common origin. One of the tracks is that of a photoelectron ejected
from an inner shell of the atom by the incident radiation. The second
track is that of an Auger electron, ejected during a radiationless
reorganization of the atom. The kinetic energy, T_P, of the photoelectron is given by

$$T_P = h\nu - E_S, \tag{1.3}$$

where E_S is the ionization energy of the atom for ejection of an
electron from the inner shell concerned and ν ($> E_S/h$) is the
frequency of the exciting radiation.

The lengths of the photoelectron tracks increase with increase
of the exciting frequency. The kinetic energy, T_A, of the Auger
electrons, on the other hand, is given by (1.2) above, and is thus
independent of the frequency of the incident radiation. By varying
this frequency Auger showed that, in the paired tracks, the length
of one partner increased with the frequency, while the length of the
other remained unchanged.

Plate I(d) shows an example of pictures obtained by Auger in
which as many as four tracks originate at the same point. Such
tracks arise from a succession of radiationless transitions in the
same atom, leaving the atom multiply-ionized in the outer levels.

Evidence for the importance of radiationless transitions in the
reorganization of an atom after inner-shell ionization came also
from the experiments of Robinson (1923) and his collaborators
(1928, 1930) on the magnetic analysis of the energy of electrons
ejected from inner shells by means of X-rays. Just as in the case of
Auger's experiment, Robinson found, in addition to photoelectrons
of energy given by (1.3) above, other groups of electrons of energy
independent of the exciting frequency, and these were interpreted
as Auger electrons.

Although the work of Auger provided the most convincing

PLATE I

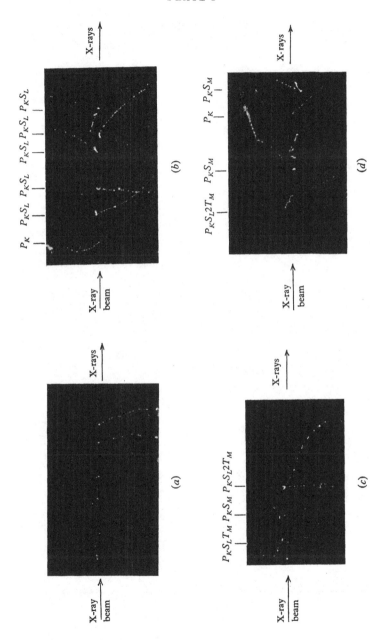

For explanation see p. xi

evidence of the occurrence of radiationless transitions, indirect evidence for processes of this kind was provided by other experiments.

Measurements of the total ionization produced by X-rays in gases were carried out by Beatty (1911) and by Barkla and Philpot (1913), using homogeneous X-radiation of different wave-lengths. They found that as the frequency of the radiation was increased through the K absorption limit, the total ionization increased sharply by two or three times. They showed that this sudden increase in ionization could not be due to the K series fluorescent radiation, since only a small part of this radiation could be absorbed in the gas of the chamber. Nor could it arise from the photoelectrons ejected from the K shell, since, for radiation of frequency very little greater than that of the K limit, these photoelectrons would have very small energies. The increase could be explained, however, by assuming the presence of Auger electrons arising from a radiationless reorganization process. Such electrons would have an appreciable range and could produce a considerable amount of ionization, independent of the quantum energy of the exciting radiation.

In the early study of X-ray processes, measurements were also made of the efficiency of production of X-ray K series radiation by fluorescence. From the known absorption coefficient of the primary radiation, the number of atoms ionized per unit time in the K shell could be calculated. Measurements of the total intensity of the K series fluorescence radiation produced then enabled the number of quanta of K series radiation emitted per K ionized atom to be measured. This ratio is called the K series fluorescence yield and is denoted by the symbol ϖ_K. Barkla and Sadler (1917) showed that for iron, ϖ_K is only about 0·3. Evidently the great majority of K-ionized iron atoms are reorganized by means of a process not involving the emission of K-series radiation.

1.2. The Auger effect and X-ray spectra

It is clear that transitions of the Auger type must play a very important role in determining intensity relations in X-ray spectra. The relative intensities of X-ray spectral lines arising from transitions involving different final levels depend markedly on the relative fluorescence yields for an atom ionized in these levels.

Further, after an Auger transition an atom may be left doubly ionized in inner shells. X-radiation from such an atom consists of 'satellite' lines. Thus the excitation of many types of X-ray satellites is markedly dependent on the Auger effect, as was first pointed out by Coster and Kronig (1935).

1.3. Should the Auger effect be considered as an internal conversion process or a radiationless transition?

So far we have referred to the Auger effect as a process of radiationless reorganization of an atom. There has been a good deal of discussion in the literature, however (e.g. Bothe, 1925), on the question of whether the process should not rather be regarded as one in which a quantum of X-radiation is first produced, and then absorbed by an electron of the atom in which it originates, before it has a chance to escape. The probability of such an absorption in the same atom would be expected to be much greater than the probability of absorption in a neighbouring atom owing to the concentration of the radiation field around the point of production.

It will be shown in subsequent sections that in the calculation of the rate of radiationless transitions it is immaterial whether the phenomenon is regarded as an internal conversion of X-radiation or the result of a direct interaction of the two electrons. In the non-relativistic limit both points of view lead to the same answer. In the relativistic theory the only method available for dealing with the problem regards the effect as an internal conversion process. If, however, a calculation is also made of the rate of emission of radiation, it is found that, owing to the direct interaction of the two electrons, the radiation rate is hardly affected by the fact that radiationless transitions are occurring at the same time. The radiation rate is almost the same as that which would be produced by a single electron in the same atomic field. In view of this it scarcely seems possible to think of the effect as, in the main, an internal conversion. The radiation rate is actually very slightly smaller than would be the case for a single electron, so that only this small part of the effect could rightly be regarded as a true internal conversion. Also there are many instances where Auger electrons are observed, even though the corresponding optical transition is forbidden by the selection rules.

1.4. Radiationless reorganization following alternative modes of inner-shell ionization

So far we have only considered the role of the Auger effect in inner-shell reorganization of atoms initially ionized by a photo-electric effect. Naturally, however, it will be of importance in inner-shell reorganization irrespective of the mode of primary ionization. In particular, Auger transitions play an important role in phenomena subsequent to inner-shell ionization, (i) by electron impact, (ii) by internal conversion of γ-rays (see § 1.5), and (iii) by nuclear capture of inner electrons.

In this last-mentioned process an electron from an inner level of an atom is absorbed by the nucleus and a neutrino is emitted. If initially the nucleus has atomic number Z, mass number A, the resulting atom has a nucleus of atomic number $Z-1$ and mass number A, and possesses a vacancy in an inner shell. The effect provides an important technique for the study of the K capture process.

1.5. Radiationless reorganization of excited atoms and molecules—auto-ionization and predissociation

Radiationless transition processes involving outer levels of atoms or molecules also play an important role under some conditions. An example is provided by the auto-ionization of helium. The energy required for the double excitation of helium to a level such as $(2s, 2p)\,^1P$, for example, considerably exceeds the ionization energy of helium. As a result an atom of helium doubly excited in this way may reorganize by a radiationless transition to a state of equal energy in which one of the electrons returns to a $1s$ level and the other is ejected from the atom. The probability of such a reorganization process, known as auto-ionization, is quite high (Kiang, Ma and Wu, 1936; Wu, 1944).

Radiationless transitions between excited states of molecules give rise to the phenomenon of predissociation in molecular spectra. For diatomic molecules this phenomenon occurs most commonly when the potential energy curves corresponding to two excited states, one of which corresponds to dissociation of the molecule, have a cross-over point. A molecule may be excited to the bound

6

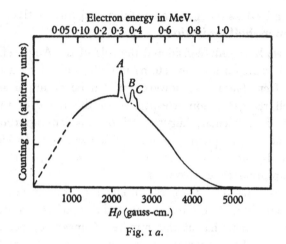

Fig. 1 *a*.

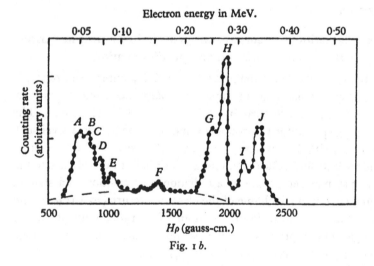

Fig. 1 *b*.

Fig. 1. Typical β-ray spectra. (*a*) β-Spectrum from ¹⁹⁸Au (2·7*d*), showing three superposed groups of homogeneous electrons (*A, B, C*) due to the internal conversion of a γ-radiation of energy 0·411 MeV. in the *K, L* and *M* shells respectively. (*b*) β-Spectrum from ¹⁹⁶Au (5·6*d*). Ten strong groups of homogeneous electrons are superposed. Groups *ABCD* are due to the Auger electrons from

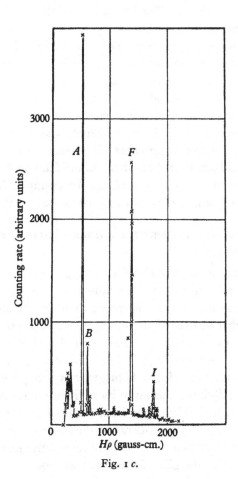

Fig. 1 c.

the processes $K \rightarrow LL$, $K \rightarrow LM$, $K \rightarrow LN$, $K \rightarrow MM$ respectively. The pairs E and F, G and H, I and J arise from the conversion of γ-radiations of energies 0·175, 0·330 and 0·358 MeV. in the K and L shells (Steffen, Huber and Humbel, 1949). (c) Electron spectrum from Th (B + C + C″), indicating the high resolution obtainable with natural radioactive sources (Flammersfeld, 1939).

8 THE AUGER EFFECT

molecular state, but at the cross-over point it may make a transition to the other state, leading to dissociation. If the lifetime of the molecule in the bound excited state is short compared with the period of molecular rotation, the rotational structure of the band spectra becomes diffuse. The broadening of the rotational lines is known as predissociation. The phenomenon of predissociation is even more common in the spectra of polyatomic molecules.

A vast field of atomic and molecular phenomena grouped under the heading of collisions of the second kind and embracing the quenching of resonance radiation, sensitized fluorescence, photo-sensitized chemical reactions and many others, also may be re-garded primarily as examples of radiationless collisions which bear a close relation to the Auger effect. The treatment of these topics, however, would take us far beyond the scope of this book. The reader is referred to other sources, such as, for example, Massey and Burhop, *Electronic and Ionic Impact Phenomena* (1952).

1.6. Radiationless nuclear transitions—internal conversion of γ-radiation

β-Ray spectra consist of electrons having a continuous range of energy from zero up to a definite upper limit. Superimposed on this continuous distribution, however, there are frequently to be found a number of sharp lines corresponding to electrons that are homogeneous in energy. Fig. 1 shows typical β-ray spectra, for the species ^{198}Au, ^{196}Au and Th$(B+C+C'')$ illustrating these features.

The interpretation of the homogeneous groups of electrons accompanying β-decay was given by Meitner (1922) and by Ellis and Skinner (1924). They arise from 'internal conversion' of the γ-radiation accompanying β-decay by the electrons of the atom of origin of the γ-radiation, the process being exactly analogous to the Auger effect. The energies of these homogeneous groups of elec-trons have been used to determine the quantum energies of the γ-radiation accompanying β-decay. In general 'internal conver-sion' is a much less probable process for γ-radiation than for X-radiation.* It has nevertheless proved important in the study of

* The process is always referred to as internal conversion of γ-rays even though it should more correctly be considered as a direct interaction between the nucleus and the outer electrons involved, just as in the Auger effect (see §2.6).

the atomic nucleus, because, as will be shown later, it provides a means for determining whether nuclear transitions giving rise to γ-radiations are dipole, quadrupole or multipole.

1.7. Production of electron pairs by internal conversion of γ-rays

In the process which results in the production of pairs of positive and negative electrons interaction occurs between an excited nucleus and an electron in a state of negative kinetic energy. Under the influence of this interaction, a transition in the nucleus, which would normally result in the emission of a quantum of γ-radiation, is associated instead with the transition of the electron to a state of positive total energy, leaving a vacancy in the distribution of electrons of negative kinetic energy. This vacancy behaves as a positron. Just as in the ordinary case of γ-ray internal conversion already discussed, this process also could be regarded as one in which a γ-ray quantum is first emitted by the nucleus and is absorbed by an electron in a state of negative kinetic energy.

1.8. Radiationless annihilation of the positron

A positron is annihilated with the production of radiation when an electron makes a transition to the vacant state of negative kinetic energy which is identified with the positron. If this annihilation process takes place in the absence of matter, conservation laws require the emission of two quanta of radiation. If it takes place in the field of a nucleus, one quantum may be emitted, excess momentum being taken up by the atom. In this latter case there is an alternative possibility, viz. the energy liberated as a result of the annihilation may be used to eject one of the electrons from the atom, the whole transition being radiationless.

1.9. Auger processes in the capture of slow negative mesons by nuclei

It is believed that the capture of a slow negative meson by a nucleus takes place in stages. First the meson is captured into a discrete level of high total and azimuthal quantum number in the field of the nucleus, then, by a series of transitions, it reaches an orbit of low total quantum number, and finally it is captured by the nucleus itself by a process similar to the K capture process for electrons.

In estimating the lifetime of the negative meson after capture into the high 'atomic' level and before capture by the nucleus it is obviously important to consider transitions without radiation as well as ordinary radiative transitions. Phenomena of this type can occur in principle for both π^- and μ^- mesons.

From the examples given in this and in following sections it will be clear that radiationless transitions of the Auger type are of importance over a very wide range of phenomena in atomic and nuclear physics.

CHAPTER II

THE THEORY OF THE AUGER EFFECT

Corresponding to the two ways of regarding the Auger effect, either as a radiationless transition due to the direct interaction of the two electrons concerned, or as an internal absorption process, we can formulate the theory along two possible lines. The theory of the Auger effect was first given by Wentzel (1927).

2.1. The non-relativistic theory of the Auger effect treated as due to the direct interaction of two electrons

In the non-relativistic approximation in the case of direct interaction one considers the two electrons in their initial states moving in the field due to the nucleus and the average field of the remainder of the electrons of the atom. The Coulomb interaction between the two electrons is then regarded as the perturbation, under the action of which a transition takes place without change of total energy to a state of the system in which one of the electrons is in a discrete state of lower energy and the other electron is raised to the continuum outside the atom.

Let $\chi_i(\mathbf{r}_1)$ and $\psi_i(\mathbf{r}_2)$ represent the initial single-electron wave functions of the two electrons concerned, in the atom in which an inner level is unoccupied. Let $\chi_f(\mathbf{r}_1)$ be the final wave function of one of the electrons in this level and $\psi_f(\mathbf{r}_2)$ the final wave function of the other electron in a state of positive total energy. We ignore for the time being the effect of the Pauli exclusion principle, which will be taken into account later.

Then the number of radiationless transitions occurring in time dt is given by first-order perturbation theory as†

$$b_n\,dt = 2\pi\hbar^{-1}\left|\int\int\chi_f^*(\mathbf{r}_1)\,\psi_f^*(\mathbf{r}_2)\,\frac{e^2}{|\mathbf{r}_1-\mathbf{r}_2|}\,\chi_i(\mathbf{r}_1)\,\psi_i(\mathbf{r}_2)\,d\mathbf{r}_1 d\mathbf{r}_2\right|^2 dt,\tag{2.1}$$

provided the continuous wave function $\psi_f^*(\mathbf{r}_2)$ is normalized to

† See, for example, Mott and Massey, *Theory of Atomic Collisions*, 2nd ed. (Oxford), 1949, p. 361.

represent one ejected electron per unit time per unit energy range. The condition for the validity of this expression is that $b\tau \ll 1$, τ being a time of the order of $1/\nu_{fi}$, where ν_{fi} is the frequency of the X-ray quantum which would have been emitted in the radiative transition $\chi_i \to \chi_f$.

2.2. The theory of the Auger effect treated as an internal absorption of radiation

To formulate the theory in the case in which the process is considered as an internal absorption of radiation, we write down the equations relativistically and obtain the non-relativistic expression by making $c \to \infty$.

During the transition process the atom is considered as a classical oscillating charge distribution of density

$$\rho_{fi} \exp\left(-2\pi i \nu_{fi} t\right) + \rho_{fi}^{*} \exp\left(2\pi i \nu_{fi} t\right), \qquad (2.2)$$

and corresponding current vector

$$\mathbf{j}_{fi} \exp\left(-2\pi i \nu_{fi} t\right) + \mathbf{j}_{fi}^{*} \exp\left(2\pi i \nu_{fi} t\right). \qquad (2.3)$$

Classical electromagnetic theory gives, for the scalar and vector potentials, A_0, $\mathbf{A}$ of the field radiated by such an oscillating charge density, the equations

$$\left. \begin{aligned} \nabla^2 A_0 - \frac{1}{c^2}\frac{\partial^2 A_0}{\partial t^2} &= -4\pi\{\rho_{fi}\exp\left(-2\pi i\nu_{fi}t\right)+\rho_{fi}^{*}\exp\left(2\pi i\nu_{fi}t\right)\}, \\ \nabla^2\mathbf{A} - \frac{1}{c^2}\frac{\partial^2\mathbf{A}}{\partial t^2} &= -\frac{4\pi}{c}\{\mathbf{j}_{fi}\exp\left(-2\pi i\nu_{fi}t\right)+\mathbf{j}_{fi}^{*}\exp\left(2\pi i\nu_{fi}t\right)\}. \end{aligned} \right\} \qquad (2.4)$$

The solution of (2.4) representing an outgoing wave only is

$$\left. \begin{aligned} A_0(\mathbf{r},t) &= a_0(\mathbf{r})\exp\left(-2\pi i\nu_{fi}t\right)+a_0^{*}(\mathbf{r})\exp\left(2\pi i\nu_{fi}t\right), \\ \mathbf{A}(\mathbf{r},t) &= \mathbf{a}(\mathbf{r})\exp\left(-2\pi i\nu_{fi}t\right)+\mathbf{a}^{*}(\mathbf{r})\exp\left(2\pi i\nu_{fi}t\right), \end{aligned} \right\} \qquad (2.5)$$

where

$$\left. \begin{aligned} a_0(\mathbf{r}) &= \int \frac{1}{|\mathbf{r}-\mathbf{r}'|}\exp\left(2\pi i\nu_{fi}\,|\mathbf{r}-\mathbf{r}'|/c\right)\rho_{fi}(\mathbf{r}')\,d\mathbf{r}', \\ \mathbf{a}(\mathbf{r}) &= c^{-1}\int \frac{1}{|\mathbf{r}-\mathbf{r}'|}\exp\left(2\pi i\nu_{fi}\,|\mathbf{r}-\mathbf{r}'|/c\right)\mathbf{j}_{fi}(\mathbf{r}')\,d\mathbf{r}'. \end{aligned} \right\} \qquad (2.6)$$

We now suppose that the field given by (2.5) and (2.6) is produced by the single-electron transition $\chi_i \to \chi_f$.

Taking χ_i, χ_f to be four-component Dirac wave functions as required in the relativistic theory, the charge and current densities are given in terms of these functions by the relations

$$\left.\begin{array}{l} \rho_{fi} = -e\chi_f^* \chi_i, \\ \mathbf{j}_{fi} = e\chi_f^* \rho_1 \boldsymbol{\sigma} \chi_i, \end{array}\right\} \tag{2.7}$$

ρ_1, $\boldsymbol{\sigma}$ being the usual matrices occurring in the Dirac theory.

The electromagnetic field given by (2.5), (2.6) and (2.7) is now considered to interact with the second electron which, under the influence of this perturbation, is ejected from the atom. The energy of interaction of an electron with a field specified by the potentials a_0, $\mathbf{a}$ is given by

$$V = -ea_0 - e\rho_1 \mathbf{a} . \boldsymbol{\sigma}. \tag{2.8}$$

The first term in (2.8) arises from the interaction of the field with the electrostatic charge of the electron, the second term from the interaction of the magnetic moment of the electron with the field.

Then in time dt the number of transitions of the electron into the final state of the continuous spectrum, ψ_f, is given by first-order perturbation theory as†

$$b_r dt = \hbar^{-2} \left| \int \psi_f^*(\mathbf{r}_2)\{-ea_0 - e\rho_1 \mathbf{a} . \boldsymbol{\sigma}\} \psi_i(\mathbf{r}_2)\, d\mathbf{r}_2 \right|^2 dt. \tag{2.9}$$

In the non-relativistic limit when $c \to \infty$ and $\mathbf{a} \to 0$

$$a_0 \to -\int \frac{e}{|\mathbf{r}-\mathbf{r}'|} \chi_f^* \chi_i\, d\mathbf{r}', \tag{2.10}$$

so that (2.9) reduces to (2.1). Thus in the non-relativistic calculation of the Auger transition rate both points of view lead to the same result.

2.3. Relativistic theory of the Auger effect

No method is available in the relativistic theory for calculating the Auger transition rate by considering the perturbation due to the direct interaction between the two particles in analogy to the non-relativistic equation (2.1). The only satisfactory development of a relativistic theory of the Auger effect involves the use of the radiation field potentials as described in the last section. This

† See Mott and Massey, loc. cit.

method of treating the perturbations due to the interaction of two electrons was first introduced by Møller (1931).

Formally, there are two ways in which the transition probability obtained on the relativistic theory, b_r, differs from the non-relativistic transition probability, b_n. In the first place the amplitude in (2.9) contains, in addition to the term in ρ_{fi} arising from the oscillating charge density, which goes over to the amplitude in (2.1) in the non-relativistic limit, also a term in $\mathbf{j}_{fi}$ which vanishes in that limit. The term in ρ_{fi} may be interpreted as arising from the direct electrostatic interaction of the two electrons, that in $\mathbf{j}_{fi}$ as arising from the direct interaction of the magnetic fields of the two electrons produced by their spins.

Again, there is the term $\exp\left(2\pi i \nu_{fi} \mid \mathbf{r} - \mathbf{r}' \mid /c\right)$ in the expression (2.6) for $a_0(\mathbf{r})$, $\mathbf{a}(\mathbf{r})$, which becomes 1 in the non-relativistic limit. This is the retardation term, and it takes account of the difference in phase of the electromagnetic wave between the two points $\mathbf{r}$, $\mathbf{r}'$, due to the finite time of propagation between these two points. In a certain sense one might perhaps consider the effect of this retardation term as constituting a specifically 'internal absorption' effect, while the expression obtained by putting this term equal to unity could be considered as constituting an effect of the direct interaction between the particles. Such a view, however, would be naïve, because, owing to the role of the electromagnetic field in the interaction between two charged particles, any theory that set out to calculate only the result of the direct interaction in giving rise to the Auger effect would necessarily introduce a retardation term.

The retardation term will only be of importance for values of $\mid \mathbf{r} - \mathbf{r}' \mid$ of the order of magnitude of the wave-length of the radiation. For most X-ray cases this means that retardation effects will be negligible because the contribution to the probability amplitude is small for values of $\mid \mathbf{r} - \mathbf{r}' \mid$ much greater than the mean separation of the electrons in the atom.

For example, in the case of the Auger effect involving two electrons in the L shell of gold when one electron makes a transition to a K shell and the other is ejected, the wave-length of the K_α radiation for gold ($K \to L$ transition) is 0·185 Å., while the radius of the L orbit for gold, from the neighbourhood of which the major contribution to the probability amplitude comes, is 0·026 Å.

For internal conversion of γ-rays, however, when the wavelength of the γ-radiation may be comparable with or less than the radius of the orbit from which the electron is ejected, the effect of retardation may be very important indeed.

2.4. The Pauli principle and the Auger effect

So far no account has been taken of the fact that in a two-electron problem the initial and final wave functions must be anti-symmetrical in their coordinates. In the non-relativistic case this is allowed for in (2.1) by replacing the product

$$\chi_i(\mathbf{r}_1)\,\psi_i(\mathbf{r}_2)$$

by $\qquad 2^{-\frac{1}{2}}\{\chi_i(\mathbf{r}_1)\,\psi_i(\mathbf{r}_2)-\chi_i(\mathbf{r}_2)\,\psi_i(\mathbf{r}_1)\}, \qquad (2.11)$

and similarly for the product $\chi_f^*(\mathbf{r}_1)\,\psi_f^*(\mathbf{r}_2)$. In these expressions the spin coordinates are now included in $\mathbf{r}_1$, $\mathbf{r}_2$.

On making this substitution the total transition probability for the Auger process involving two electrons initially in states represented by wave functions χ_i, ψ_i becomes

$$b_n'dt = 2\pi\hbar^{-1}\,|\,f_1^n - f_2^n\,|^2\,dt, \qquad (2.12)$$

where f_1^n is the amplitude given in equation (2.1) and

$$f_2^n = \iint \chi_f^*(\mathbf{r}_2)\,\psi_f^*(\mathbf{r}_1)\,\frac{e^2}{|\,\mathbf{r}_1-\mathbf{r}_2\,|}\,\chi_i(\mathbf{r}_1)\,\psi_i(\mathbf{r}_2)\,dr_1 dr_2. \qquad (2.13)$$

f_1^n is the probability amplitude corresponding to the transition

$$\chi_i \to \chi_f, \quad \psi_i \to \psi_f;$$

f_2^n is that corresponding to the 'exchange' transition

$$\chi_i \to \psi_f, \quad \psi_i \to \chi_f.$$

In both transitions the ejected electrons have the same energy, and there is no other way of distinguishing between the two processes experimentally. The transition probability given by (2.12) represents their combined effect.

In the relativistic theory it is more difficult to allow correctly for the Pauli principle because no complete relativistic quantum theory of two-electron processes has yet been developed. It is clear, however, that in the non-relativistic limit the relativistic formula must reduce to (2.12). One would expect, then, the correct relativistic expression to be

$$b_r' \, dt = 2\pi\hbar^{-1} \, | \, f_1^r - f_2^r \, |^2 \, dt, \qquad (2.14)$$

where f_1^r, f_2^r are the relativistic probability amplitudes corresponding to f_1^n, f_2^n.

2.5. Calculation of the fluorescence yield

Equations (2.12) or (2.14) enable one to calculate the relative probabilities of occurrence of different Auger processes. These calculations can be compared with the results of experiments like those of Robinson and his collaborators (see § 1.1). However, most experiments on the Auger effect have aimed at the determination of the fluorescence yield, ϖ, for atoms ionized in a given inner shell. To calculate this one needs to know, in addition to the total transition rate for all radiationless transitions to the vacant level, the rate of ionization in the inner level. In equilibrium, the number of atoms per unit time that have the vacant inner shell filled by all processes, including radiation, will equal the number per unit time ionized in the given inner shell.

Let the suffix p refer to the vacant inner level of an atom and m, n to two occupied levels of smaller ionization energy. Let a_n^p be the rate at which the p level is filled by radiative transitions from the level n and b_{nm}^p the rate at which it is filled by Auger transitions arising from the interaction of two electrons in the n and m levels.

Then the fluorescence yield for the p level is given by

$$\sum_n a_n^p / \{ \sum_n \sum_m b_{nm}^p + \sum_n a_n^p \}, \qquad (2.15)$$

where in the case of $\sum_n a_n^p$ the summation is over all levels having energy greater than that of the p level, and in the case of $\sum\sum b_{nm}^p$ the summation is over all pairs of levels for which an Auger process filling the p level is energetically possible.

Because they are so much more probable than other transitions only electric dipole transitions need be considered in the calculation of the a's. The number of such radiative transitions occurring in time dt is

$$a \, dt = \{ 64\pi^4 \nu^3 \, | \, \mathbf{M} \, |^2 / 3hc^3 \} \, dt, \qquad (2.16)$$

where ν is the frequency of the radiation emitted and $\mathbf{M}$, the matrix element of the electric dipole moment, is given by the expression

$$e\int \chi_f^* \mathbf{r}\chi_i\, dr. \qquad (2.17)$$

This expression is valid in both the non-relativistic and relativistic theories, although in the latter case the initial and final electronic wave functions, χ_i, χ_f, are four-component Dirac wave functions.†

2.6. The theory of the internal conversion of γ-rays

The calculation of the transition rate for internal conversion of γ-radiation follows lines similar to those used for the calculation of the fluorescence yield. The question of indistinguishability of particles does not arise, however, because the nuclear particles concerned in the transition giving rise to the radiation cannot be electrons. On the other hand, there is no precise knowledge of the wave functions of these particles or of the type of transition involved.

Calculation of the internal conversion rate for γ-rays has therefore been carried out assuming a radiator (electric or magnetic dipole or quadrupole, etc.) located at the centre of the nucleus. The appropriate scalar and vector potentials corresponding to the assumed type of radiation are substituted in equation (2.9) to give the transition probability for internal conversion of the radiation accompanied by the electronic transition $\psi_i \rightarrow \psi_f$.

For example, for electric dipole radiation of frequency ν one has to substitute in (2.9)

$$\left.\begin{aligned} a_0 &= Br^{-1}\exp{(iqr)}\,(1+i/qr)\cos\theta, \\ \mathbf{a} &= \mathbf{i}a_x + \mathbf{j}a_y + \mathbf{k}a_z, \end{aligned}\right\} \qquad (2.18)$$

where $\qquad a_x = a_y = 0, \quad a_z = Br^{-1}\exp{(iqr)}, \quad q = 2\pi\nu/c.$

Similarly, for electric quadrupole radiation

$$\left.\begin{aligned} a_0 &= Cr^{-1}\exp{(iqr)}\{1+(3\cos^2\theta-1)(1+3i/qr-3/q^2r^2)\}, \\ a_x &= a_y = 0, \\ a_z &= 3Cr^{-1}\exp{(iqr)}\,(1+i/qr)\cos\theta, \end{aligned}\right\} \qquad (2.19)$$

† The use of the expressions (2.16), (2.17) implies that the X-ray transition rate is not affected by the occurrence of the Auger transitions. This is almost, but not quite, true (see §2.6).

and for magnetic dipole radiation the similar potentials are

$$\left.\begin{aligned}
a_0 &= a_z = 0,\\
a_x &= Kr^{-1}\exp(iqr)(1+i/qr)\sin\theta\sin\phi,\\
a_y &= -Kr^{-1}\exp(iqr)(1+i/qr)\sin\theta\cos\phi.
\end{aligned}\right\} \qquad (2.20)$$

In these expressions B, C, K are undetermined constants.

Calculations of the radiative transition rate are made by calculating from (2.5) the rate at which energy is radiated corresponding to these field potentials and dividing by $h\nu$. Thus p, the mean radiative transition rate, is given by

$$\left.\begin{aligned}
p &= 16\pi^2 B^2\nu/3hc \quad \text{(electric dipole),}\\
p &= 48\pi^2 C^2\nu/5hc \quad \text{(electric quadrupole),}\\
p &= 16\pi^2 K^2\nu/3hc \quad \text{(magnetic dipole).}
\end{aligned}\right\} \qquad (2.21)$$

The internal conversion coefficient α of a γ-radiation by an electron in a given atomic level is defined as the ratio of the number of conversion electrons ejected from that level to the number of γ-ray quanta emitted per unit time. A difficulty arises in the calculation of α, since it is not obvious whether the number of γ-ray quanta emitted per excited nucleus per unit time should be given by p or by $p-b$, where b, given by (2.9), is the number of internal conversion electrons emitted per unit time per excited nucleus. The answer to this question was supplied by Taylor and Mott (1933), who showed that in the presence of the extra-nuclear electrons the total rate of energy radiation by the nuclear multipole is not given by $ph\nu$ but is somewhat greater—in fact, nearly equal to $(p+b)h\nu$. The internal conversion coefficient α is then given, very nearly, by

$$\alpha = b/p, \qquad (2.22)$$

and not by $b/(p-b)$, as had been supposed prior to their work.†

Taylor and Mott (1933) used a perturbation calculation to make allowance for the interaction between the electrons and the nuclear particles undergoing the transition. They considered a system

† The definition of α is further elaborated in §5.1. The definition given here is that usually adopted in the contemporary literature. Some writers, however, still use for α the ratio of the number of conversion electrons to the total number of relevant nuclear transitions (with or without radiation). If all the internal conversion occurs in a single level this is practically equivalent to writing $b/(p+b)$ instead of b/p for α in (2.23).

comprising the nucleus and the particular atomic electron ejected in the process.

Let $\mathbf{R}$, $\mathbf{r}$ be respectively the coordinates of the nuclear particle and the atomic electron, η and $-\epsilon$ their respective charges and

$$\dot{\Phi}_i\,(=\phi_i(\mathbf{R},\mathbf{r})\exp(-iW^it/\hbar)),\quad \Phi_f\,(=\phi_f(\mathbf{R},\mathbf{r})\exp(-iW^ft/\hbar)) \quad (2.23)$$

their initial and final wave functions. W^i, W^f are here the initial and final total energies of the system.

Then the radiating charge density corresponding to the transition is given by the real part of

$$\left\{\eta\int\Phi_f^*\Phi_i\,d\mathbf{r}-\epsilon\int\Phi_f^*\Phi_i\,d\mathbf{R}\right\}. \quad (2.24)$$

To the zero-order approximation the initial wave function is

$$\Phi_i^0=\Psi_n(\mathbf{R})\,\psi_0(\mathbf{r})\exp\{-i(W_n+E_0)\,t/\hbar\}, \quad (2.25)$$

where $\Psi_n(\mathbf{R})\exp\{-iW_nt/\hbar\}$ is the wave function of the excited particle of total energy W_n in the nucleus and $\psi_0(\mathbf{r})\exp\{-iE_0t/\hbar\}$ the wave function of the electron of total energy E_0.

In the first-order approximation account is taken of the interaction between the nuclear particle and the electron. Suppose there is only one other nuclear state available of energy W_0 ($<W_n$), and that the interaction V between the two particles is 'switched on' at time $t=0$. Then at time t, perturbation theory will give for the wave function of the total system

$$\Phi_i^1=c_0(t)\Psi_n(\mathbf{R})\,\psi_0(\mathbf{r})\exp\{-i(W_n+E_0)\,t/\hbar\}$$
$$+\Psi_0(\mathbf{R})\,\chi(\mathbf{r},t)\exp\{-iW_0t/\hbar\}, \quad (2.26)$$

instead of (2.25). In this equation $\chi(\mathbf{r},t)$ represents a wave corresponding to the ejected electron diverging from the atom. $c_0(t)$ is a function that will differ little from unity provided the perturbation is small so that $pt\ll 1$. Equation (2.26) simply expresses the fact that owing to the perturbation, the initial state of the system is not stationary, but is a linear superposition of all the possible states of the system of the same total energy. In the case here supposed there are only two such available states.

The function Φ_f, to be substituted in equation (2.24), represents the state in which both electron and nuclear particle are in their lowest energy states, i.e.

$$\Phi_f=\Psi_0(\mathbf{R})\,\psi_0(\mathbf{r})\exp\{-i(W_0+E_0)\,t/\hbar\}. \quad (2.27)$$

Putting (2.26) and (2.27) in (2.24), we obtain for the radiating charge density for the transition

$$\eta \Psi_0^*(\mathbf{R}) \Psi_n(\mathbf{R}) \exp(-2\pi i\nu t) - \epsilon \psi_0^*(\mathbf{r}) \chi(\mathbf{r}, t) \exp(iE_0 t/\hbar). \quad (2.28)$$

The first term in (2.28) is just the radiating charge density for the nuclear transition unperturbed by the presence of the electron. This is the radiating charge density that gives rise to the radiation field with potentials given by (2.18), (2.19) or (2.20). If the corresponding transition is an electric dipole the constant B of (2.19) is

$$B = (2\pi\eta\nu/c) \int \Psi_0^* z \Psi_n d\mathbf{R}. \quad (2.29)$$

The second term in (2.28) represents the radiating charge density due to the presence of the electron. Taylor and Mott calculated the change in the radiating field potentials arising from this second term.

They found that the effect of this term is to modify the expressions (2.18), (2.19), (2.20) for the electromagnetic potentials of the field due to the nuclear transition. In the case of an electric quadrupole transition, for instance, the field quantities given by (2.19) had to be replaced by potentials of the form

$$\left. \begin{aligned}
a_0(\mathbf{r}) &= 3C\{1 + \lambda_2 + \lambda_{-3}\} r^{-1} \exp(iqr) \cos^2\theta, \\
a_x(\mathbf{r}) &= -Cr^{-1} \exp(iqr)\{3\eta_2(5\cos^2\theta - 1)/2 + \eta_{-3}\} \sin\theta \sin\phi, \\
a_y(\mathbf{r}) &= Cr^{-1} \exp(iqr)\{3\eta_2(5\cos^2\theta - 1)/2 + \eta_{-3}\} \sin\theta \cos\phi, \\
a_z(\mathbf{r}) &= 3C\{1 + \lambda_2 + \lambda_{-3}\} r^{-1} \exp(iqr) \cos\theta,
\end{aligned} \right\} \quad (2.30)$$

where λ_2, λ_{-3}, η_2, η_{-3} are small quantities of the order $e^2/\hbar c$.

The rate of radiation from such a modified field, instead of being given by (2.21) is now given by

$$(48\pi^2\nu^2C^2/5c) J,$$

where
$$J = |1 + \lambda_2 + \lambda_{-3}|^2 + \tfrac{10}{7}|\eta_2|^2 + \tfrac{5}{9}|\eta_{-3}|^2, \quad (2.31)$$

and this represents actual radiation, irrespective of whether radiationless transition processes also occur.

Numerical evaluation of J showed that it was, in fact, less than unity, but only by a quantity of the order $e^2/\hbar c$.[†]

This calculation showed then that, as a result of the direct interaction between the nuclear particle and the atomic electron, the transition rate for the nuclear particle is increased by an amount almost, but not quite, equal to the rate for ejection of the electron from the atom, so that it is correct to put α nearly equal to b/p as in (2.22).

[†] Tralli and Goertzel (1951) have derived a similar result while Coish (1951) has calculated the correction to the rate of electron emission.

This calculation of Taylor and Mott enables us now to give a full answer to the question of whether Auger effects arise from internal conversion or from radiationless transitions due to the direct interaction of electrons. Almost the whole of the ejected electrons arises from direct interaction. However, a small fraction (of the order of magnitude $e^2/\hbar c$) arises from a true internal absorption process.

As a check on their method Taylor and Mott applied it to the case in which the perturbing field was that of a light wave coming from outside the atom. In this case it was found that the diminution in intensity of the radiation was exactly equal to the energy per cm.^{2} per sec. required to eject the electrons. This is what would be expected when there is no possibility of direct interaction between the atom and the source of radiation.

In the case of the relativistic Auger effect the expression (2.18) used for the calculation of the number of radiative transitions should be slightly modified because similar arguments to those used for the γ-rays will apply also in this case, when the interaction between the two atomic electrons is considered in place of the interaction between the electron and nuclear particle. The true radiation rate will be a little smaller than (2.17), (2.18) by a fraction of the order of $e^2/\hbar c$, which is negligible from the point of view of comparison between experiment and theory.

2.7. Internal conversion effects arising from forbidden transitions

Consider an atom ionized in a K shell. Suppose the atom re-organizes by a transition involving the two electrons in the L_1 shell, one of these electrons filling the vacancy in the K shell and the other being ejected. It is usual to denote such a transition by the symbol $K \to L_1 L_1$, the letters referring to the inner shell ionization in the initial and final states. Then the expression (2.1) for such a process gives a finite result. If, however, one were to regard this process as arising from an internal conversion, the radiation would correspond to the transition $2s \to 1s$ which is always forbidden. The Auger effect occurring in this case must clearly be regarded as arising from a direct interaction between electrons.

Effects of this kind were first noted by Fowler (1930) in the interpretation of γ-ray internal conversion. He interpreted a homogeneous group of electrons of energy $1\cdot33$ MeV. in the β-ray spectrum of Ra C' as arising from a transition between two known states of the Ra C' nucleus. But the γ-ray of energy $1\cdot42$ MeV. corresponding to such a transition had never been observed. Fowler postulated then that the two states involved in the transition corresponded to a nuclear angular momentum $j = 0$, so that the corresponding radiative transition between them was totally forbidden. Nevertheless, owing to the direct interaction between the nuclear particle undergoing transition and the atomic electron, nuclear transitions could occur and would result in the ejection of the electron from the atom.

We return to a more detailed discussion of the internal conversion of γ-radiation in Chapter v.

CHAPTER III

EXPERIMENTAL STUDY OF THE
AUGER EFFECT

The fluorescence yield corresponding to ionization in a specified inner shell $(K, L, ...)$ is defined as the fraction $(\varpi_K, \varpi_L, ...)$ of the atoms so ionized that reorganize with the emission of radiation.

A number of different methods is available for the production of the primary ionization. As the name 'fluorescence yield' suggests, ϖ has most commonly been determined employing the method of fluorescent excitation. A beam of X-radiation is passed through the specimen being studied, and the rate of inner-shell ionization either measured or calculated from known absorption coefficients. In principle inner-shell ionization by electron impact could also be employed, but this method has not so far been used for the direct measurement of the fluorescence yield, since the calculation of the rate of ionization is made difficult owing to the complicated effects arising from electron scattering in the target.

Measurements of the fluorescence yield have more recently been made for atoms which have been ionized in an inner shell following nuclear transformation. The process of K capture results in ionization of this type, and Steffen, Huber and Humbel (1949), for example, have measured the K fluorescence yield for two isotopes of platinum produced by the K capture process in two gold isotopes. The internal conversion of γ-radiation also gives rise to inner-shell ionization, and several investigators have studied the fluorescence yield in such cases (Arnoult, 1939; Flammersfeld, 1939; Kinsey, 1948 a, b).

Knowing the rate of inner-shell ionization, the fluorescence yield can be calculated if either the rate of emission of characteristic radiation or of Auger electrons can be directly measured or deduced from other observations.

The most direct method for measuring the fraction of reorganizations which occur with the emission of Auger electrons consists in counting the number of paired tracks appearing in cloud-chamber photographs of the type studied by Auger and comparing this with the number of ordinary photoelectron tracks. The rate of emission of Auger electrons has also been measured by means of Geiger counters, the energy of the electrons being determined either by magnetic analysis or by absorption methods (Kinsey, 1948 b).

Another method of determining the fluorescence yield from the Auger electron emission rate consists in the measurement of the sudden change in ionization per absorbed quantum produced by a beam of X-rays in a gas when the frequency of the X-radiation is varied through a critical frequency.

If the reorganization process is studied by measuring the intensity of the characteristic radiation rather than that of the Auger electrons, the most direct method consists in the use of an ionization chamber to compare the power in the fluorescent beam with the power in the exciting beam. In a variant of this method, the intensity of the radiation produced has been measured by a counter of known efficiency (Kinsey, 1948 a).

In addition to these direct methods, the fluorescence yield has been estimated indirectly in some cases from experimental determinations of the total width of X-ray levels and theoretical estimates of the partial width arising from radiative transitions.

Recently, the development by Curran of proportional counting techniques for electron detection (Curran, Angus and Cockcroft, 1949) has provided another method for the determination of fluorescence yields for gases. Photographic emulsion techniques have also been used (Germain, 1950).

3.1. The cloud-chamber method

The cloud-chamber method has been employed for the measurement of the fluorescence yield by Auger (1925, 1926), Locher (1932), Martin and his collaborators (1935, 1937) and Bower (1936). The method is clearly applicable only to materials available as gases or vapours. It is a statistical method, and therefore the observation of a considerable number of tracks is needed to obtain high accuracy, and in doubtful cases there may be scope for personal errors in the selection of tracks. Confusion may also be produced by ionization of the water vapour or other condensant always present in the cloud chamber. The method is very simple in principle. A narrow

PLATE II

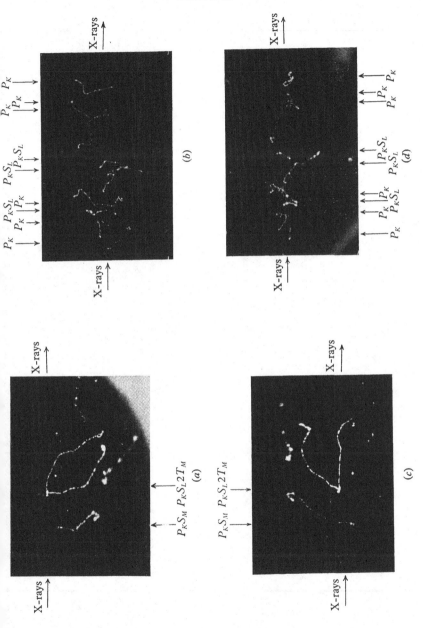

For explanation see p. xi

beam of homogeneous X-radiation is passed through the expansion chamber which contains the gas being studied, and tracks of the photoelectrons ejected from the various atomic levels are observed, as well as the paired tracks arising from the Auger effect (see Plates I and II). As the frequency of the exciting radiation is varied the length of the tracks produced by the photoelectrons varies, but the length of the tracks produced by Auger electrons remains constant.

The length of the tracks due to Auger electrons will increase with the atomic number of the gas in which they are produced. For example, consider an Auger effect produced as a result of a re-organization process involving two L electrons, one of which goes over to a vacant K level, the other being ejected. Then typical approximate values of the energy of the ejected electron ($\simeq E_K - 2E_L^*$ where E_K, E_L are respectively the K and L ionization energies) for elements of widely different atomic number are given in Table I.

TABLE I

Element	Atomic number	$E_K - 2E_L \simeq$ energy of ejected electron (keV.)
A	18	3·14
Kr	36	10·7
Xe	54	24·15
Hg	80	55·4

For elements of low atomic number therefore the tracks due to the Auger electrons are very short, even for the reorganization of atoms ionized in a K shell. For example, Plate I shows typical Auger tracks obtained in argon. The Auger tracks appear only as short spurs at the origins of the photoelectron tracks. To increase the length of the Auger tracks in such cases and thus make detection easier, it is usual to dilute the argon with hydrogen.

In the experiments of Martin, Bower and Laby (1935), for example, a mixture of 10% argon and 90% hydrogen was employed. Owing to its low atomic number, very few photoelectrons are produced from the hydrogen. It is necessary, however, to make

---

* The true ejected electron energy is given by $E_K - E_L - E_L'$, where E_L' is the L shell ionization energy of a K-ionized atom (see §4.6).

a correction for the contribution from the oxygen atoms of the water vapour present, which give rise to photoelectron tracks indistinguishable from those arising from argon, but even this correction, in general, amounts to a small fraction of 1%.

In experiments with elements of higher atomic number the Auger electrons are easier to distinguish. In the limit the tracks may become so long, however, that an undesirably high proportion end in the walls of the chamber, making their identification again uncertain. In addition, many of the tracks now are multiple owing to successive Auger processes occurring in the reorganization following K shell ionization. For example, in their work on xenon, both Auger (1926) and Martin and Eggleston (1937) found many cases in which three Auger electrons, in addition to the original photoelectron, originated at a common point. Plate II contains pictures obtained by Martin and Eggleston in xenon and krypton showing examples in which the photoelectron alone is ejected, others in which it is accompanied by an Auger electron from the L shell, and another in which it is accompanied by an L electron together with two M electrons, evidently Auger electrons ejected from the M shell in the further reorganization of the atom.

If symbols $P, S, T, \ldots$ are used to denote primary photoelectron, secondary and tertiary electrons, respectively, emitted in the reorganization process, and the suffixes denote the shell in which the electrons originate, processes of the type P_K, $P_K S_M$, $P_K S_L$, $P_K S_L T_M$, $P_K S_L 2T_M$ were observed in xenon.* In the process $P_K S_M$ the energy of the Auger electron was approximately $E_L - 2E_M$. Tracks such as these must arise from atoms ionized in the K shell initially, which first reorganize by the emission of a quantum of K_α radiation. The vacancy in the L shell is then filled in an Auger reorganization resulting in the emission of an electron. Clearly, from the point of view of K shell reorganization, tracks of type $P_K S_M$ must be regarded as due to transitions with emission of radiation.

Tracks of type $P_K S_L$, $P_K S_L T_M$, $P_K S_L 2T_M$, where the energy of the ejected L electron is $E_K - 2E_L$) and of the M electron is

---

* Thus $P_K S_M$ refers to an event in which a primary K electron and a secondary M electron are ejected. $P_K S_L 2T_M$ is an event in which a primary K electron, a secondary L electron and two tertiary M electrons are ejected.

$E_L - 2E_M$), follow the filling of a K shell vacancy by Auger transitions. No cases of Auger reorganizations giving secondary electrons of energy $E_K - 2E_M$ or $E_K - E_L - E_M$ were recognized; the straggling of range would make it difficult to distinguish these from the more frequent $P_K S_L$. After radiationless reorganization, involving two L electrons, the atom is left doubly ionized in the L shell. Tracks of type $P_K S_L T_M$ are then obtained when one of these L shell vacancies is filled by an Auger transition, and the other by a transition involving the emission of radiation, and tracks of type $P_K S_L 2T_M$ when both L shell vacancies are filled by radiationless transitions. In these experiments the K fluorescence yield is given by the ratio

$$\frac{\text{Number of } P_K + P_K S_M \text{ events}}{\text{Number of } P_K + P_K S_M + P_K S_L + P_K S_L T_M + P_K S_L 2T_M \text{ events}}.$$

In the case of a light element such as argon it is not possible to detect M Auger electrons, nor to distinguish photoelectron tracks due to the ejection of electrons from the K shell from those due to the ejection of electrons from other shells. A correction has thus to be made for the fraction of the observed photoelectrons that arise from outer shells. If P is the total number of observed single photoelectron tracks and A the observed number of Auger tracks, the K fluorescence yield is given not by $P/(A+P)$ but by

$$\{PJ_K/(A+P) - 1\}/(J_K - 1), \tag{3.1}$$

where J_K is the ratio of the total number of exciting quanta absorbed to the number absorbed in shells other than the K shell.*

Since this method of determining the fluorescence yield is statistical the probable error will depend on the total number of tracks observed. If tracks from n ionized atoms are observed and of these m reorganize in a radiationless process the fluorescence yield is $1 - m/n$, and the probable error in the determination of m/n is $0.6475\{m(n-m)/n^3\}^{\frac{1}{2}}$.

The cloud-chamber method is probably the most accurate available for the measurement of the fluorescence yield in cases like

---

* J_K is sometimes called the 'K jump', and is taken to be the ratio of the X-ray absorption coefficients on the high-frequency and low-frequency sides of the K discontinuity. This is, of course, only correct when the frequency of the radiation is sufficiently small and scattering makes no important contribution to the absorption coefficient.

those of Xe and Kr in which the photoelectron tracks arising from ejection of electrons from the K shell can be distinguished from those originating in the L shell, because under such circumstances it is not necessary to rely on other measurements of X-ray absorption coefficients. In the case of the measurement of the L series fluorescence yield, however, and of the K series fluorescence yield for light elements, difficulty arises from the uncertainty in the actual value of J_K (or of J_L in the L series case).

3.2. Use of proportional counters for the measurement of the fluorescence yield

The development of the proportional counter for electron detection by Curran and his co-workers (Curran, Angus and A. L. Cockcroft, 1949) has made possible another method, closely related to the cloud-chamber method for the measurement of the fluorescent yield.

The tube counters used by Curran were of the conventional self-quenching type of effective length about 30 cm., diameter about 3 cm., with a tungsten wire 0·01 cm. diameter as anode, and filled with a mixture of methane and a suitable counting gas. In a typical counter the filling was a mixture of argon and methane at pressures of 70 and 10 cm. Hg respectively. When operating in the proportional region the amplification of the ionization produced in the active volume of the counter is of the order of 100. Since the energy required to produce a pair of ions in the gas filling the counter is about 30 eV., an electron of energy 3 keV. will produce at the counter wire a pulse corresponding to approximately 10^4 electron charges. This pulse is then amplified by means of a suitable linear amplifier and its height measured by a pulse amplitude sorter.

By using electrons of well-known energy from suitable radioactive electron emitters Curran showed that the output pulse amplitude was proportional to the energy of the electron to within about 2 %.* The possibility of using proportional counters for the measurement of the fluorescence yield arises from the fact that when an X-ray quantum is absorbed by the K shell of the counter gas and the atom reorganizes by means of an Auger transition, the full energy E of the incident X-radiation is dissipated in the counter.

---

* Further information about the performance of these counters has been given by Hanna, Kirkwood and Pontecorvo (1949).

When the atom reorganizes with the emission of K series X-radiation, however, a large proportion of the radiation will escape from the counter provided the pressure in it is not too high. In this case an energy $E - E_K^0$ is dissipated in the counter, where E_K^0 is the

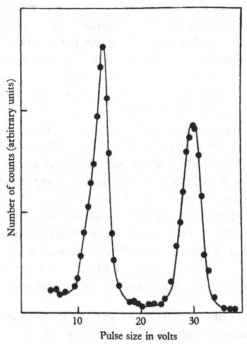

Fig. 2. Pulse size distribution for a krypton-filled proportional counter in a magnetic field of 4300 gauss. The exciting radiation was the 24·2 keV. X-radiation from a ^{113}Sn source. The peak at lower energy corresponds to the ejection of a photoelectron alone, that at higher energy to a photoelectron and an Auger electron.

energy of the K series radiation. Thus two peaks are observed in the pulse-size distribution, and the K series fluorescence yield can be determined from their relative intensities.

Correction has to be made for the reabsorption of the K series radiation in the counter gas, but this correction can be kept down to about 5 % by use of a suitable gas pressure. It can then be estimated with sufficient accuracy using known absorption data. Just as in the cloud-chamber measurements a correction has also to be made for the absorption of the incident radiation in the $L, M, N, \ldots$ shells. To allow for this effect the true fluorescence yield is given by (3.1) above, where now P is the number of pulses of energy $E - E_K^0$ and A the number of energy E.

A rough value of the K fluorescence yield for argon of 0·06 was obtained by Curran, Angus and A. L. Cockcroft (1949), using this method in reasonable agreement with values obtained by Auger and by Martin *et al.* using a cloud-chamber method (see Table III).

Difficulty is experienced in attempting to extend this method to heavier elements owing to the possibility of escape of the photo-electrons and Auger electrons from the counter. To overcome this, West and Rothwell (1950) have operated proportional counters in magnetic fields of strength about 5000 gauss, sufficient to keep all but an insignificant fraction of the ejected electrons inside the counter. They showed that the magnetic field made very little difference to the linearity of the gas amplification of the counter. As sources of X-radiation to produce the primary radiation the K-capture isotopes ^{103}Pd and ^{113}Sn were used. The energies of the K radiations from these isotopes are 20·2 and 24·2 keV. respectively. Fig. 2 shows a pulse-size distribution obtained in a krypton-filled counter from a ^{113}Sn source with magnetic field of 4300 gauss. There is clearly no difficulty in distinguishing the peaks arising from the Auger and fluorescent radiation process.

3.3. Use of photographic emulsion techniques for the measurement of the fluorescence yield

The fluorescence yield for Po (84) has been measured recently by Germain (1950) using electron-sensitive photographic plates. $^{211}_{84}$Po is formed by K capture from $^{211}_{85}$At, with a half-life of 7·5 hr., but then decays (half-life 5×10^{-3} sec.) to $^{207}_{82}$Pb with emission of an α-particle. A photographic plate was soaked with a solution containing astatine which was prepared by bombarding bismuth with α-particles in the Berkeley cyclotron. The number of astatine decays prior to development of the plate could be measured by counting the number of ^{211}Po α-particles produced. By counting the number of these α-particles with Auger electrons of energy 59 keV. at the start of their track, it was possible to estimate the proportion of atoms that had reorganized by an Auger process.[*]

---

[*] Other examples of the measurement of ϖ when the inner ionization arises as a result of nuclear transformation are discussed in §3.5.

3.4. Direct measurement of the fluorescence yield

Many investigations of the fluorescence yield have been carried out by measuring directly the efficiency of production of fluorescent radiation from atoms ionized in an inner shell by a beam of primary X-radiation. From a knowledge of the intensity of the primary radiation and its absorption coefficient in the target material, the rate of inner-shell ionization can be calculated. In order to carry out this calculation it is essential to know the spectral composition of the primary beam, and clearly the calculation is made much easier if this is homogeneous in frequency.

To ensure a definite composition of the beam three methods are available. Crystalline diffraction may be employed or, if greater incident intensity is required, a suitable spectral range may be isolated by the use of appropriate absorbing materials (Ross, 1926). Another method uses as primary radiation the fluorescent radiation produced by the incidence of an intense beam of X-radiation on a suitable radiator. Characteristic radiations of definite frequency without any continuous background are produced in this way.

In addition to the absorption coefficient of the incident radiation in the target material this method requires also a knowledge of the absorption coefficient of the fluorescent radiation in the target and of the quantity J_K (or J_L) introduced above. The method involves the accurate comparison of the intensity of two X-ray beams differing in wave-length, and its successful use requires full realization of the factors governing the accurate measurement of X-ray intensity. An understanding of these factors has not always been evident. However, much careful work has been carried out using this method. It is suitable for the study of materials available in solid form.

A typical apparatus is that used by Stephenson (1937) shown in fig. 3. X-radiation, nearly homogeneous in energy, was obtained by using fluorescent radiation from a radiator R_1 placed near the window of an X-ray tube. A beam of this X-radiation, collimated by a series of circular holes, was allowed to fall on the radiator R_2, whose fluorescence yield was being studied, the surface of the radiator making an angle of $45°$ with the incident beam. Fluorescent radiation from R_2 entered the ionization chamber which was placed in the position A, and the resulting ionization current was measured. With the ionization chamber at B, and the radiator R_2 removed, the direct beam entered the chamber. By comparing the ionization currents with the chamber in the two positions, the fluorescence yield of the material of the radiator could be calculated.

In experiments of this type the power of the radiation in the primary beam which enters the ionization chamber in the position

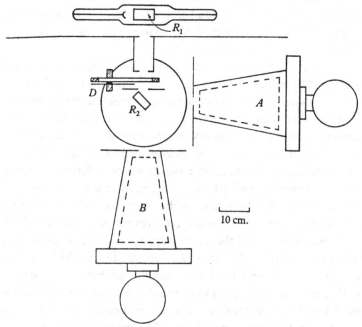

Fig. 3. Stephenson's apparatus for measuring ϖ.

B is generally many hundreds of times that of the fluorescent beam received in position A. In Stephenson's arrangement a rotating disk D with an adjustable radial slot was therefore used to reduce the power in the primary beam, so making it comparable with that of the fluorescent beam entering the chamber in position A.

Let X-radiation of intensity I and frequency ν be incident on the target. The number of quanta absorbed per unit time in a layer δx at a depth x in the radiator R_2 is

$$(Is/h\nu)\exp(-\mu x \operatorname{cosec}\theta)\tau \operatorname{cosec}\theta\,\delta x, \tag{3.2}$$

where θ is the glancing angle of incidence of the beam, s its area of cross-section, and μ, τ are respectively the total and photoelectric absorption coefficients for the incident radiation. The number of atoms, δn_i, of the layer δx ionized in the K shell is obtained by multiplying this expression by $(J_K - 1)/J_K$.

Let ϖ_K be the K series fluorescence yield and ν_α, ν_β, ... the frequencies of the various K series lines. Then the number of quanta of frequency ν_η radiated from this layer per unit time is $\varpi_K z_\eta \delta n_i$, z_η being the ratio of the number of quanta of frequency ν_η to the total number of quanta radiated.

If the ionization chamber be placed to receive radiation from R_2 at a glancing angle of emergence θ', the total energy in the radiation of frequency ν_η and target absorption coefficient μ_η entering the ionization chamber in unit time from the layer of thickness δx at depth x is

$$(\Omega/4\pi) h\nu_\eta \varpi_K z_\eta \exp\left(-\mu_\eta x \csc\theta'\right) \delta n_i$$
$$= (\Omega s/4\pi)(I/\nu)(1 - 1/J_K) \nu_\eta z_\eta \varpi_\eta$$
$$\times \exp\left\{-(\mu\csc\theta + \mu_\eta \csc\theta')x\right\}\tau \csc\theta \, \delta x,$$

where Ω is the solid angle subtended at the radiator by the entrance slit of the ionization chamber and isotropic emission is assumed.

Integrating over all thickness, for an 'infinitely thick' target, the total X-ray energy of frequency ν_η entering the chamber per unit time in position A is

$$(\Omega s/4\pi)(I/\nu)(1 - 1/J_K) \nu_\eta z_\eta \varpi_\eta \tau \csc\theta/(\mu\csc\theta + \mu_\eta \csc\theta').$$

If the incident radiation is composed of a number of discrete frequencies of absorption coefficient μ_i, and if r_i is the fraction of the total intensity in the ith component, the total X-ray energy of all frequencies entering the chamber per unit time in position A is

$$(\Omega s/4\pi)(1 - 1/J_K) \varpi_K I \sum_i \sum_\eta \{r_i z_\eta \nu_\eta \tau_i \csc\theta/\nu_i(\mu_i \csc\theta + \mu_\eta \csc\theta')\}.*$$

$$(3.3)$$

In the position B, on the other hand, the total X-ray energy incident per unit time is sI, so that the ratio of the ionization currents in the chamber in the two positions is

$$\varpi_K(\Omega/4\pi)(1 - 1/J_K)\sum_i \sum_\eta \{r_i \tau_i z_\eta \nu_\eta \csc\theta/\nu_i(\mu_i \csc\theta + \mu_\eta \csc\theta')\}.$$

$$(3.4)$$

Equation (3.4) shows that if the ionization current ratio is measured the K fluorescence yield ϖ_K can be calculated from the

---

* It is assumed the target behaves as a true 'thick target' for all the radiations concerned.

geometrical constants of the apparatus provided the composition of the incident and fluorescent beams and the appropriate absorption data are known.

It should be noted that the above analysis assumes that all the fluorescent radiation entering the ionization chamber in the position A is K series radiation. L radiation and other soft radiations can be excluded by differential absorption.

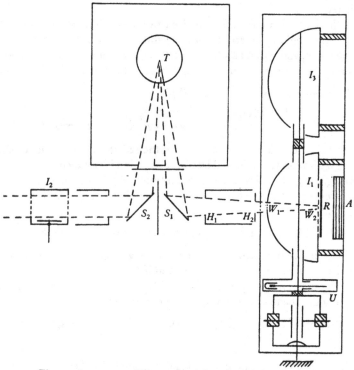

Fig. 4. Apparatus of Küstner and Arends for measuring ϖ.

Various modifications of the procedure just described have been used by different investigators. Fig. 4 shows the arrangement used by Küstner and Arends (1935).

X-rays from the tungsten target T of a commercial tube were incident on the two secondary radiators S_1 and S_2. Fluorescent radiation from S_2 entered the ionization chamber I_2, the ionization current being used to monitor the intensity of the source. Fluorescent radiation from S_1 (referred to subsequently as the primary radiation), after passing through a suitable filter, was collimated by the holes H_1, H_2 and entered the ionization chamber I_1 through a window W_1 of cellophane 0·01 mm. thick, coated with graphite. The chamber I_1 was in the form of a hemisphere of radius 20 cm. The primary radiation, after leaving the chamber through another graphite-coated cellophane window W_2, was incident on a sheet R of the material whose fluorescence yield was being measured. The radiators R were chosen of sufficient thickness to ensure that further increase of the thickness produced no appreciable increase of the fluorescent radiation returning to I_1.

The chamber I_1 was enclosed in a lead box. To avoid error from secondary radiation from the lead when R was not in position, an absorber A was placed in front of the lead in the region where the primary radiation would strike it. The absorber A consisted of a series of layers of 1·1 mm. tin, 0·5 mm. copper, 0·5 mm. aluminium and 0·32 mm. cellophane.

The purpose of this 'sandwich' was to absorb the quanta and photoelectrons produced in the lead by absorbers of lower atomic number. In this way, in the end, neither quanta nor photoelectrons could return to I_1.

The radiator R could be moved out of the way to determine the ionization produced in I_1 by the primary beam alone. A second similar chamber I_3 placed alongside I_1 was connected so as to compensate for changes in background effect.*

The ionization in I_1 was measured with and without the radiator R in position for a number of different fluorescent materials S_1. In this way the increase in X-ray intensity in I_1 due to the presence of R could be measured for different wave-lengths of the primary radiation.

Fig. 5 (a) and (b) shows, for a number of different radiators, the ratio i_s/i_p between the additional ionization current i_s due to R and the current i_p due to the primary beam for different values of $1/\lambda$, the wave number of the primary beam. For heavy radiators the increase of ionization current due to fluorescence as each of the

---

* Background fluctuations were further compensated by connecting in the small ionization chamber U, containing a uranium layer whose effective surface could be varied.

three L shells is excited is shown in fig. 4(b). Since only a finite number of frequencies of primary radiation could be used, it is clear, however, that there is considerable uncertainty in the precise shape of the curves near the L discontinuities. The method is much more precise for measurements of the increase of ionization

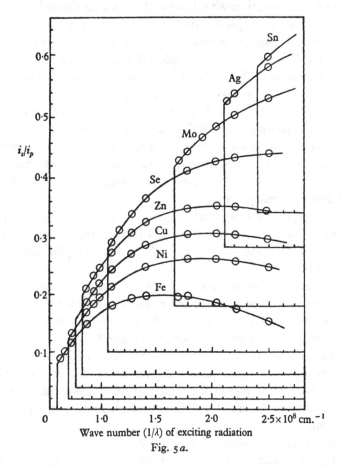

Fig. 5 a.

on passing through the K discontinuity (fig. 5 (a)). From fig. 5 (b) the part of the increase of i_s/i_p due to fluorescent radiation from each of the L shells may be determined and the fluorescence yield calculated along lines similar to those outlined above. The determinations of fluorescence yield are most likely to be accurate for the L_{III} edge which occurs at the lowest frequency.

Lay (1934), who investigated fluorescence yields following L as well as K ionization, used a photographic method for comparing the incident and fluorescent X-ray intensities, but this method is probably subject to greater inaccuracies than is the ionization chamber method.

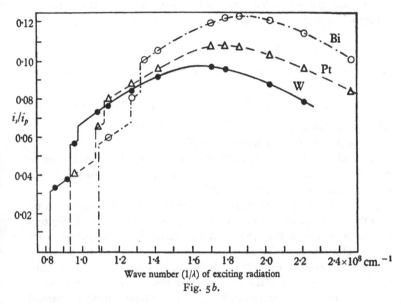

Fig. 5. Variation, with reciprocal wave-length of exciting radiation, of the ratio i_s/i_p of the secondary to primary ionization current in the measurement of the fluorescence yield. (*a*) Measurements of Arends (1935) of the K fluorescence yields of Fe, Ni, Cu, Zn, Se, Mo, Ag and Sn. The curves are displaced vertically to prevent overlap. (*b*) Measurements of Küstner and Arends (1935) of the L fluorescence yields of W, Pt, Bi.

3.5. Estimation of fluorescence yield from the change of ionization by X-rays in a gas at a critical frequency

This method was effectively first used by Martin (1927) to deduce ϖ_K for Se (34), Br (35) and I (53) from early work by Barkla and Philpot (1913) and Beatty (1911). Stockmeyer (1932) also applied it for determining ϖ_K for Br (35).

Consider a parallel beam of X-rays of intensity I_ν and frequency ν ($\nu > \nu_K$, the frequency of the K absorption edge), passing axially through an ionization chamber. Let μ be the total absorption coefficient, and τ_K, τ_L the photoelectric absorption coefficients corresponding to K and L

shell absorption respectively. Then, assuming $\tau_M, \tau_N \ldots$ can be neglected, the number of photoelectrons ejected per unit area per unit time in a layer δx of gas distant x from the entrance is

$$I_\nu \exp(-\mu x)(\tau_K + \tau_L)\,\delta x/h\nu.$$

Of these the photoelectrons ejected from the L shell produce an ionization current of amount

$$a I_\nu \exp(-\mu x)\tau_L(\nu - \nu_L)\,\delta x/\nu,$$

while that produced by the K photoelectrons, including the Auger electrons, is

$$a I_\nu \exp(-\mu x)\tau_K\{(\nu - \nu_K) + (1 - \varpi_K)(\nu_K - 2\nu_L)\}\,\delta x/\nu, \tag{3.5}$$

where a is constant.

We now make the following assumptions:

(1) Owing to their comparatively small energy, L series fluorescent quanta and Auger electrons ejected in the process of filling L shell vacancies produce negligible ionization.

(2) The fraction of the K series fluorescent quanta absorbed in the chamber is small, so that the ionization produced by it can be ignored in comparison with that produced by Auger electrons ejected in the process of filling K shell vacancies.

(3) The energy required to produce a pair of ions in the gas is independent of the energy of the photoelectron producing the ionization.

Neglecting ν_L in comparison with ν, ν_K the total ionization current produced in a chamber of length l is

$$a I_\nu \{\nu(\tau_K + \tau_L) - \varpi_K \nu_K \tau_K\}\{1 - \exp(-\mu l)\}/\mu \nu. \tag{3.6}$$

Let μ_1 and μ_2 be the total absorption coefficients on the short and long wave-length sides of the K discontinuity for the gas contained in the ionization chamber.

Then, for equal radiation intensities, the ratio R of the ionization current in the chamber when the frequency of the exciting radiation is just on the short wave-length side of the K discontinuity to the current when the frequency is just on the long wave-length side is ($\nu_1, \nu_2 \to \nu$)

$$R = \mu_2\{1 - \exp(-\mu_1 l)\}(\tau_K + \tau_L - \varpi_K \tau_K)/\mu_1 \tau_L\{1 - \exp(-\mu_2 l)\}. \tag{3.7}$$

Approximately, $\mu_2/\mu_1 = \tau_L/(\tau_K + \tau_L)$, so that

$$R = \{1 - \varpi_K \tau_K/(\tau_K + \tau_L)\}\{1 - \exp(-\mu_1 l)\}/\{1 - \exp(-\mu_2 l)\}. \tag{3.8}$$

Using this equation ϖ_K can be calculated from a measurement of R and absorption data.

3.6. Measurement of fluorescence yield using calibrated Geiger-Müller counters

Atoms may be ionized in inner shells as a result of the process of internal conversion of γ-radiation. If the intensity of the characteristic radiation or of the Auger electrons emitted in the subsequent reorganization can be measured, the fluorescence yield can be calculated. This method is very useful because it can be applied to obtain approximate values in cases in which the methods described above are not very accurate. It is suitable, for example, when $\varpi \sim 1$ as for the K yield of heavy elements.

Similarly, K or L ionized atoms may be produced by the capture of K or L orbital electrons by nuclei. This process, which produces a nucleus one lower in atomic number than the initial nucleus, is energetically possible provided

$$M(Z, A) > M(Z - 1, A) + E_K(Z - 1), \qquad (3.9)$$

where $M(Z, A)$ is the mass of the atom of atomic number Z, mass number A and $E_K(Z)$ its K shell ionization energy expressed in mass units.

An arrangement which has been used by Kinsey (1948b) to determine L series fluorescence yields for Th B, Th C and Ra D atoms is shown in fig. 6.

The radioactive material being investigated, present in the form of a very thin layer on thin aluminium, was placed at O. The L series radiation from it was detected by the Geiger-Müller counter C containing a mixture of argon, xenon and alcohol. The beam S_2 was well collimated by the series of holes, H, and electrons emitted from the source were prevented from reaching the counter by the magnetic field supplied by a small electromagnet NS. The counter C was screened from external radiation by the lead shield B. Nickel-absorbing foils placed in the beam enabled the quantum energy of the radiation reaching the counter to be determined. In this way it was shown that the radiation recorded by C consisted mainly of the L series radiation of the material being studied, on which was superposed a small amount of harder γ-radiation for which a correction could be made without difficulty.

The source itself was surrounded by a steel cylinder D which screened it from the magnetic field. A small hole K in this cylinder allowed α- or β-particles to pass through a thin window into another counter P, working either in the proportional region or as a Geiger-Müller counter.

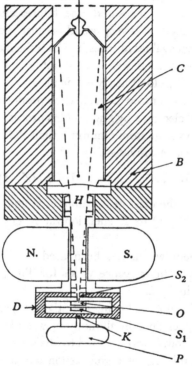

Fig. 6. Apparatus used for the measurement of L-series fluorescence yields by Kinsey (1948b).

The fluorescence yield was calculated from the ratio ϵ of the counting rate in C due to L-radiation to the counting rate in P due to α- or β-particles from the source.

Two methods were used to measure ϵ. In one of these the two rates were measured independently. In the other, coincidences were observed between the α-particles and the pulses in the counter C and the ratio ϵ of the coincidence rate to the total counting rate of P determined. To separate the L-radiation from the harder γ-radiation the difference in coincidence rates was measured with and without a nickel filter (30 mg./cm.^{2}) between the source and the counter C.

Let Ω_C and Ω_P be respectively the solid angles for collection by the counters C and P, R an end-correction for the counter C representing the mean increase of counter efficiency by absorption and re-radiation at the end of the counter, z_1, z_2 and z_3 the number of conversions in the three L subshells per disintegration, ϖ_1, ϖ_2 and ϖ_3 the fluorescence yields for these three subshells, η_ν the efficiency of detection of the counter C for L-radiation of frequency ν, and r_ν the fraction of all the quanta originating from an initial state of ionization in a particular subshell, that are emitted in the line of this frequency.

Then the ratio of the two counting rates measured in the experiment becomes

$$\epsilon = (\Omega_C/\Omega_P)\,R \sum_\nu z\varpi r_\nu \eta_\nu, \qquad (3.10)$$

where in the summation over all the L-series radiations the values of z, ϖ, r_ν are in each case those appropriate to the initial state of L-ionization leading to the emission of the line of frequency ν.

η_ν can be calculated* knowing the absorption coefficient of the radiation in the counter gas, R can be estimated, r_ν, ν are known from other experiments, and hence a relation involving the fluorescence yields ϖ_1, ϖ_2 and ϖ_3 for the three L shells can be obtained.

Of course this method cannot be used to determine the ϖ's explicitly, but it was used by Kinsey to confirm estimates of these quantities for RaD and ThB made by other means. The latter experiments were more difficult to interpret because they were made with a Th$(B+C)$ source, and the effects of L-radiation from ThB, ThC, and ThC″ had to be separated out.

A counter method has been applied to obtain more definite information about the fluorescence yield for platinum K-radiation by Steffen, Huber and Humbel (1949). A target of platinum was bombarded by protons of energy 7 MeV. in a cyclotron, and gold isotopes ^{194}Au, ^{195}Au, ^{196}Au and ^{198}Au were produced by a p-n reaction. The isotopes ^{194}Au and ^{196}Au decay to ^{194}Pt and ^{196}Pt respectively by K-capture processes with half-lives of 39 hr. and

---

* The correction for absorption of the radiation in the air between the source and the counter is also included in this factor η_ν.

5·6 days respectively. In the experiment the K fluorescence yield following this decay was measured.

The method will be illustrated by describing in detail the procedure in the case of ^{194}Pt. Similar sets of measurements were carried out for the other isotopes, the effects due to the different isotopes being separated out by making the measurements at different times after the exposure. Most of the radiation immediately after the exposure arises from ^{194}Pt.

The energies of the γ-radiations emitted from ^{194}Pt were determined by measuring the absorption of the Compton electrons

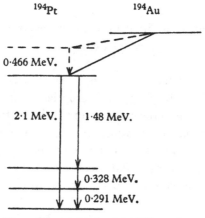

Fig. 7. Energy levels of the ^{194}Pt nucleus.

produced by them in an aluminium plate. A coincidence counting system was used for this purpose.

Five distinct γ-ray components were detected with quantum energy 2·1, 1·48, 0·466,* 0·328, and 0·291 MeV. respectively, and the emission of these components was interpreted in terms of the set of energy levels shown in fig. 7.

Most of the ^{194}Pt nuclei were formed in a state of excitation of 2·1 MeV. above the ground-level and then fell to the ground-level either (1) by the emission of three γ-ray quanta in cascade, or (2) by emission of a single quantum. The three-quantum process took place in about 70 % of all cases.

* The radiation of quantum energy 0·466 MeV. was weak.

By using a γ-ray counter of which the counting efficiency for each γ-ray separately could be estimated, it was possible to measure the relative rates N_1, N_2 at which the processes (1) and (2) were occurring in the ^{194}Pt nucleus.

A lead plate 2 cm. thick was then placed between the source and the counter to eliminate all but the two hardest components of the γ-radiation.

If the counter efficiencies for these two components are η_1, η_2 respectively, and if μ_1, μ_2 are the corresponding absorption coefficients in the lead block of thickness d the γ-ray counting rate Z_γ is given by

$$Z_\gamma = N_1 \Omega_\gamma \eta_1 \exp\left(-\mu_1 d\right) + N_2 \Omega_\gamma \eta_2 \exp\left(-\mu_2 d\right), \quad (3.11)$$

where Ω_γ is the solid angle expressed as a fraction of a complete solid angle of the γ-ray beam. The ratio N_1/N_2 having been determined as above, N_1, N_2 may be obtained separately from (3.11).

Coincidences were next measured between the hard γ-rays and electrons ejected by internal conversion in the inner shells of the platinum atom. Since internal conversion of the 2·1 MeV. ray is very small, the measured coincidence rate K_1 is given by

$$K_1 = N_1 \Omega_\gamma \Omega_\beta \alpha_1 \eta_1 \exp\left(-\mu_1 d\right), \quad (3.12)$$

where Ω_β is the fractional solid angle of collection of the electron counter and α_1 the total internal conversion coefficient for the three quanta emitted in cascade in process 1. Since N_1 is known, the internal conversion coefficient α_1 could be calculated.

Finally, a magnetic spectrograph was employed to derive the energy spectrum of all the electrons ejected from ^{194}Pt, using a thin-walled counter as detector. Electron groups were observed corresponding to the conversion of each of the γ-rays in the K, L and M shells, and also to the Auger electrons arising from the conversion of K series radiation. The relative intensities of these different groups of electrons were measured.

If α_1, α_2 are respectively the total internal conversion coefficients corresponding to the processes (1) and (2), α_{1K}, α_{2K} the portions of these that correspond to internal conversion in the K shell and I_1, I_2 and I_A the relative numbers of electrons produced by internal conversion of the γ-radiation in these processes and the number of Auger electrons respectively, the total number of K-ionized atoms produced by all processes is $N_1(1+\alpha_{1K})+N_2(1+\alpha_{2K})$, and the number of Auger electrons ejected is $\alpha N_1 I_A/I_1$, so that the K fluorescence yield is

$$1-\alpha_1 N_1 I_A/\{N_1(1+\alpha_{1K})+N_2(1+\alpha_{2K})\} I_1. \qquad (3.13)$$

3.7. Results of the measurement of fluorescence yield and their comparison with calculation:

(i) *The K-shell fluorescence yield*

The magnitude of the fluorescence yield is independent of the frequency of the radiation exciting the fluorescence. This result must hold if the interpretation given above of the Auger processes is correct.

Typical results indicating this independence are shown in Table II.

TABLE II

Radiator	Exciting radiation	Wave-length (Å.)	Fluor-escence yield	Observer
A	Cu K_α	1·5	0·074	Martin, Bower and Laby (1935)
A	Mo K	0·5	0·080	Martin, Bower and Laby (1935)
A	W K	0·2	0·076	Martin, Bower and Laby (1935)
Zr	Sn K	0·480	0·69	Stephenson (1937)
Zr	Ba K	0·378	0·69	Stephenson (1937)
Mo	Sn K	0·480	0·73	Stephenson (1937)
Mo	Ba K	0·378	0·74	Stephenson (1937)

The results of the measurements that have been made of the K fluorescence yield by the methods described above are collected in Tables III and IV. Wide discrepancies are evident between the results of different workers. In view of the difficulties of the measurements this is perhaps understandable. Some of the discrepancies are undoubtedly due to differences in the X-ray absorption data, particularly in the values of the K jump, J_K, assumed by different workers. Thus in the work of Locher, using the cloud-chamber method, the absorption data seemed so uncertain that two widely different sets of values for the fluorescent yield were given, corresponding to the use of different values of J_K. However, even allowing for uncertainty in absorption data, there is still considerable disagreement between the observational data of different

TABLE III. K-series fluorescence yields, ϖ_K for elements of small and medium atomic number

Element	(1) Cloud-chamber method			(2) Direct measurement of fluorescence yield										(3) Change of ionization at K edge		(4) Proportional counter method		(5) Geiger-counter method	(6) Theoretical value
	Auger (1925)	Locher (1932)	Martin and co-workers (1935, 1937)	Arends (1935)	Backhurst (1936)	Balderstone (1926)	Berkey (1934)	Compton (1929)	Haas (1932)	Harms (1927)	Lay (1934)	Martin (1927)	Stephenson (1937)	Barkla and Beatty (Martin, 1927)	Stockmeyer (1932)	Curran and co-workers (1949)	West and Rothwell (1950)	Bergström and Thulin (1950 b)	Non-relativistic theory
O (8)		0·023																	0·0045
Ne (10)		0·028																	0·011
Mg (12)									0·013										0·021
Si (14)									0·038										0·037
S (16)									0·083										0·056
Cl (17)	0·07	0·128	0·077						0·108		0·15								0·068
A (18)																0·06			0·081
Ca (20)									0·150		0·207	0·32							0·119
Cr (24)							0·38		0·263		0·265								0·22
Fe (26)				0·302		0·375	0·39				0·343								0·28
Co (27)							0·43												0·31
Ni (28)				0·364		0·445	0·45	0·37		0·28	0·436	0·39	0·385						0·34
Cu (29)	0·51		0·53	0·401		0·495	0·53			0·38		0·45	0·41	0·68					0·37
Zn (30)				0·450		0·57	0·55	0·55		0·40	0·476	0·52	0·48	0·68					0·40
As (33)								0·56											0·49
Se (34)				0·550			0·72			0·52	0·585		0·575				0·66		0·52
Br (35)								0·68							0·56				0·54
Kr (36)							0·79			0·62									0·57
Sr (38)										0·73									0·62
Zr (40)				0·724		0·97							0·69						0·67
Mo (42)						0·88							0·735						0·71
Rh (45)				0·795									0·77						0·76
Pd (46)					0·785		0·72						0·81						0·78
Ag (47)					0·801		0·70						0·79						0·79
Cd (48)				0·825	0·835		0·66						0·81						0·81
Sn (50)					0·838		0·64												0·83
Sb (51)					0·846		0·59												0·84
Te (52)					0·855														0·85
I (53)					0·862														0·86
Xe (54)	0·70		0·78		0·872									0·88			0·81	0·89	0·87
Ba (56)					0·900														0·89

authors. Thus, for argon, Table V shows a comparison of the ratio $P/(A+P)$ obtained from cloud-chamber data in different experiments.

TABLE IV. *K-series fluorescence yields ϖ_K for elements of high atomic number**

Element	Observer	Theoretical value	
		Non-relativistic	Relativistic
Pt (78) 0·94 Th C (83) 0·93 Th C (83) 0·97 Po (84) 0·894	Steffen, Huber and Humbel (1949) Kinsey (Ellis), (1948 b) Kinsey (Flammersfeld), (1948 b) Germain (1950)	0·97	0·94

* Germain used a photographic emulsion technique. The other measurements were made with calibrated counters.

TABLE V

	No. of tracks observed	$P/(A+P)$
Auger	169	0·162
Locher	1950	0·240
Martin, Bower and Laby	2328	0·169

It seems evident that the discrepancy between Locher's work and that of the other authors must arise from the difficulties of identification of the Auger electron tracks which are very short in the case of argon. The cloud-chamber method should give much better results for krypton, although for xenon, where still shorter wavelength exciting radiation has to be used, confusion may be caused by the Compton recoils.

The experiments which measure the fluorescence yield directly are subject to even greater error than are cloud-chamber measurements, unless they are carried out with great care. As an example of the magnitude of the error to be expected for low atomic number elements, in a region where X-ray absorption data are not very well known, Haas (1932) has estimated the uncertainty of his results as from 15 to 20%. The measurements are probably more accurate in the region of atomic numbers 25–40, but become more difficult again for higher atomic numbers, where, owing to the much shorter wave-length required for the exciting radiation, the determination of its intensity becomes increasingly uncertain.

Care has to be taken also to ensure that the exciting radiation is reasonably homogeneous in frequency or, if this is not possible, then its spectral composition must be accurately known. For example, the presence of an unsuspected soft component in the exciting radiation, of too long a wave-length to produce K shell ionization in the material being studied, would lead to an underestimation of the fluorescence yield. Backhurst (1936) has suggested that some such effect might explain the low values obtained for the fluorescence yield for elements of atomic number above 42 by Berkey (1934). Berkey's results, showing an actual decrease of ϖ_K with increasing atomic number, have not been confirmed by any other workers and appear most unlikely to be correct, on theoretical grounds.

Balderstone's work (1926) has been criticized by Compton (1929) on the ground that he did not correctly allow for the scattering of the exciting radiation in the ionization chamber. Harms (1927), in calculating ϖ_K, assumed that the average energy required by a fast electron to produce an ion pair in a gas depends on the energy of the electron. Harm's results have since been corrected by Compton to allow for the fact that this assumption is now known to be incorrect. The results as corrected by Compton are given in Table III.

The estimate of ϖ_K for Se, Br and I made by Martin (1927), using the change of ionization at the K edge, as already stated, were based on old measurements by Barkla and Philpot (1913) and Beatty (1911) and are probably not very accurate.

The results given in Tables III and IV are shown graphically in fig. 8.

Theoretically, calculations of the fluorescence yield have been made by Burhop (1935) and Pincherle (1935 a) using a non-relativistic theory, and by Massey and Burhop (1936 a) using a relativistic theory. In these calculations hydrogen-like single-electron wave functions were used, the effective nuclear charge being that given by the application of Slater's rules.

The expressions for the non-relativistic Auger transition rate, obtained by integrating equation (2.1), are complicated but show that this transition rate is almost independent of the atomic number of the material being investigated. On the other hand, the transition rate for the production of radiation, obtained by integrating equation (2.17), increases approximately with the fourth power of the

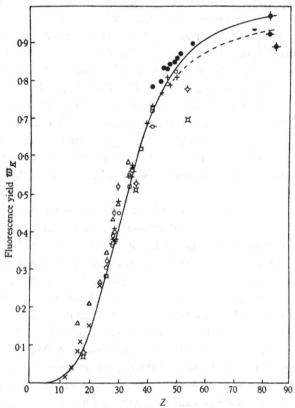

Fig. 8. Measured variation of ϖ_K with atomic number Z, compared with the calculated curves. —— non-relativistic theory; ---- relativistic theory. Experimental points: ⊄ Auger (1926); -◇- Martin et al. (1935, 1937); ○ Arends (1935); ● Backhurst (1936); -○- Compton (1930); × Haas (1932); □ Harms (1927); △ Lay (1934); ◊ Martin (1927); + Stephenson (1937); ▲ Stockmeyer (1932); ▼ Steffen, Huber and Humbel (1949); -●- Ellis (1933a); ◆ Flammersfeld (1939); -◆- Germain (1950).

atomic number. Thus equation (2.16) for the fluorescence yield for an element of atomic number Z may be written

$$(1 + a_K Z^{-4})^{-1}, \tag{3.14}$$

where a_K is a constant. The experimental results are well represented by an expression of this form. This is shown by fig. 9, where

$\varpi_K/(1-\varpi_K)$ is plotted against Z^4 for the most reliable of the experimental values. If relation (3.14) is followed this plot should be linear. The straight line which best represents the experimental points corresponds to a value of a_K of $1\cdot12\times10^6$.

The expected value of a_K can be deduced from the theoretical calculations. The non-relativistic calculations of Burhop (1935)

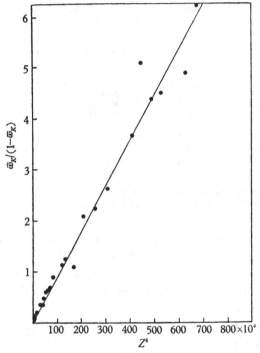

Fig. 9. The variation of $\varpi_K/(1-\varpi_K)$ with Z^4. The experimental points shown are the averaged results of Arends, Backhurst, Compton, Haas, Harms, Lay, Martin, Stephenson and Auger.

gave accurately only the Auger transition probabilities corresponding to the interaction of the L-shell electrons. The calculations of Pincherle (1935 a), however, were more extensive. Using Burhop's calculations to obtain the transition rates corresponding to Auger transitions arising from the interactions of L-shell electrons, and those of Pincherle to estimate the relative importance of other types of Auger transition, Table VI shows the transition rates to be expected for both Auger and radiative transitions in silver.

These calculations give for the fluorescence yield of silver the value $\varpi_K = 0.793$, which would correspond to a value of a_K of 1.27×10^6, in fair agreement with the value 1.12×10^6 deduced above from the experimental results.

TABLE VI

(1) Auger transitions		(2) Radiative transitions	
Transition	No. of transitions per atomic unit of time* ($\times 10^{-3}$).	Transition	No. of transitions per atomic unit of time ($\times 10^{-3}$)
$K \to LL$	28·3	$K \to L$	155
$K \to LM$	16·3	$K \to M$	37
$K \to MM$	2·1	$K \to M$	5
$K \to LN$	3·9		
$K \to MN$	0·7		
$K \to NN$	0·1		
Total	51·4	Total	197

* The atomic unit of time used here is the time of description of the lowest Bohr orbit in hydrogen divided by 2π. It is 2.42×10^{-17} sec.

It should be remembered however that, for elements of lower atomic number, the M shell does not start to fill until $Z = 11$, and the N shell starts to be filled at $Z = 19$ but is not completed until $Z = 46$. It is therefore of interest to calculate what the constant a would become:

(1) if only interactions involving L electrons are possible, as is the case for $Z \leqslant 10$,

(2) if only interactions involving L and M electrons are possible as for $10 < Z \leqslant 18$.

For interactions involving L electrons only a_K comes out to be 0.90×10^6, while for interactions involving L and M electrons it is 1.19×10^6.

Taking account of the incompleteness of electronic shells for elements of low atomic number the best non-relativistic theoretical values of the fluorescent yield would appear to be those given in the last column of Table III.

The calculations of Massey and Burhop (1936a) indicate that when account is taken of relativistic effects by the methods outlined in Chapter II, important differences in the theoretical fluorescence yield are to be expected for elements of high atomic number. The calculations involved in the relativistic theory are very lengthy, and they have so far been carried out only for the case where one electron is in the L_I subshell and the other electron in an L_I, L_{II} or L_{III} subshell. These calculations show, however, that for gold the quantity $1 - \varpi_K$, sometimes called the K-series internal conversion coefficient, is probably about twice as great, when calculated on the relativistic theory, as would be expected from the non-relativistic theory.* Of interest therefore are the measured fluorescence yields for Pt (78), Th C (83) and Po (84) in Table IV. If expression (3.14) continued to hold, the expected value of ϖ for $Z = 80$ would come out to be 0·97, while the relativistic calculations would appear to indicate a somewhat lower value, possibly in the neighbourhood of 0·94. The measurements of Steffen, Huber and Humbel (1949), which are probably the best of those available for large Z, indicate a value of 0·94 for Pt (78). For Th C (83) the position is indefinite, since the measurements of Ellis (1933 a) indicate a value $\varpi = 0·93$, while those of Flammersfeld (1939) give $\varpi = 0·97$. The recent measurements of Germain (1950) for Po (84) give $\varpi = 0·894$.

The theoretical non-relativistic and estimated relativistic values of ϖ_K are shown in fig. 8.

(ii) *The L-shell fluorescence yield*

The experimental information on the L-shell fluorescence yield is less complete than that for the K shell. The fluorescence yields for the subshells L_I, L_{II} and L_{III} will in general be different. Some observers have measured the fluorescence yields for the three L subshells separately, and the results obtained depend only on the atom studied and not on the method of excitation. Often, however, only the mean yield for the three L subshells has been measured. Care has to be exercised in interpreting such measurements, since the relative excitation of the three subshells depends critically

---

* This assumes that the other transitions that have not been calculated (particularly $K \to L_{II, III} L_{II, III}$) show a relativistic effect similar to that for the transitions $K \to L_I L_{II, III}$. Since it is not possible to say definitely what the effect on the uncalculated transitions is likely to be, the above estimate of the change in ϖ_K must be regarded as provisional.

on the mode of excitation. There are four ways in which vacancies may be produced in the L shells, and for each of these the relative ionization probabilities ϕ_{L_I}, ϕ_{L_II}, ϕ_{L_III} in the L_I, L_II and L_III subshells are different.

For fluorescent excitation the relative ionization probability depends on the frequency. For frequencies some way above the critical frequencies $\phi_{L_\mathrm{I}}:\phi_{L_\mathrm{II}}:\phi_{L_\mathrm{III}} = 1:2:3$ very roughly.

On the other hand, for excitation by electron impact the ratios $\phi_{L_\mathrm{I}}:\phi_{L_\mathrm{II}}:\phi_{L_\mathrm{III}}$ are more nearly $1:1:2$.

In cases where the L-shell ionization is caused by internal conversion of γ-radiation the corresponding ratios may be nearly $100:10:1$ and depend on the quantum energy of the radiation.

The L-shell ionization may also arise from reorganization of an atom ionized in the K shell. In this case the transition $K \rightarrow L_\mathrm{I}$ cannot occur in radiative transitions, and the ratios $\phi_{L_\mathrm{I}}:\phi_{L_\mathrm{II}}:\phi_{L_\mathrm{III}}$ are nearly $0:1:2$ for heavy elements where the K shell is filled mostly by radiative transitions.

Clearly, then, for each method of producing L-shell ionization the mean L-series fluorescence yield will be different.

Auger (1926) and Bower (1936) have used the cloud-chamber method to measure the mean L-series fluorescence yield for the fluorescent excitation of xenon and krypton. The actual ratios of the number of Auger to photoelectron tracks observed in the experiments of these two investigators agree well, but there is some uncertainty about the value to be taken for the L jump which determines the fraction of the observed photoelectron tracks which originate in the L shell.

Direct measurement of the fluorescence yield has been carried out for a number of elements by Lay (1934), using a photographic method to compare the intensities of the incident and fluorescent beams.

Measurements of the fluorescence yield for the three subshells separately have been made by Küstner and Arends (1935), and for the L_III shell alone by Stephenson (1937). By using incident radiation of a suitable frequency, conditions can be obtained in which ionization takes place (i) in the L_III subshell alone, (ii) in the L_II and L_III subshells, and (iii) in all three L subshells. By measuring the fluorescence yield by methods similar to those described in

§3.4 for each of these cases, the fluorescence yields for the three L subshells separately may be obtained.

Calibrated counters have been used by Kinsey (1948 b) to determine the L fluorescence yields of Th C, Th C″, Th D and Ra E.

An indirect method of estimating the L fluorescence yield for elements between Ta (73) and U (92) is due to Kinsey (1948 a). It is based on the best available measurements of the total width of X-ray line and absorption edges for heavy elements (Richtmyer, Barnes and Ramberg, 1934; Williams, 1934; Bearden and Snyder, 1941; Cooper, 1942; Coster and de Langen, 1947) and on the values of the expected contribution to the radiation width calculated relativistically by Massey and Burhop (1936 b). The total width, Γ, of a state of inner-shell ionization of an atom is proportional to the total probability per unit time of transitions in which the state of ionization is destroyed. Similarly, the radiation with Γ_R is proportional to the total transition probability per unit time for radiative transitions. Thus the X-ray fluorescence yield is given by Γ_R/Γ.

The width of an X-ray emission line is equal to the sum of the widths of the initial and final states. Thus the difference in width of two lines with a common initial state is equal to the difference in width of the two final states. If the width of one of the final states is determined from the shape of the corresponding absorption edge, the width of the others may then be calculated from line width measurements.

Care has to be taken in the estimation of fluorescence yields for the three L subshells for elements in which the L-shell vacancy may be shifted from one subshell to another by radiationless transitions. Such transitions will be discussed in some detail in Chapter IV. In certain cases they may occur with a very high probability.* Let $n_{L_{\mathrm{I}}}$, $n_{L_{\mathrm{II}}}$, $n_{L_{\mathrm{III}}}$ be the number of atoms ionized in the three L subshells, respectively, and let α_{rs} be the probability that a vacancy in the level L_r is transferred to L_s $(s > r)$ by a non-radiative transition, then in the calculation of the fluorescence yield the number of vacancies in the three subshells must be taken as $n_{L_{\mathrm{I}}}$, $n_{L_{\mathrm{II}}} + \alpha_{12} n_{L_{\mathrm{I}}}$ and $n_{L_{\mathrm{III}}} + \alpha_{13} n_{L_{\mathrm{I}}} + \alpha_{23}(n_{L_{\mathrm{II}}} + \alpha_{12} n_{L_{\mathrm{I}}})$ respectively.

* The shifting of the excitation between L subshells may also occur through radiative transitions but the probability of this is very small.

Results of the measurements of L fluorescence yields are shown in Table VII. There seems to be reasonable agreement for the L_{III} subshell between the measurements of Küstner and Arends (1935) and of Stephenson (1937) on the one hand, and the values estimated by Kinsey from line widths on the other. For the L_I and L_{II} subshells, however, there is violent disagreement between the measurements of Küstner and Arends and the estimates of Kinsey (1948 a). Thus Küstner and Arends find the L_I-fluorescence yield larger than the L_{II} or L_{III} yields, while the line width estimates show just the opposite behaviour. Owing to the possible occurrence of Coster-Kronig (see Chapter IV) transitions for large Z the estimates of Kinsey appear much more reasonable.

For the L_{II} subshell the measurements of Küstner and Arends show an actual decrease of $\varpi_{L_{II}}$ with increase of Z. This is quite contrary to theoretical expectation, and once again the estimates of Kinsey are likely to be more reliable.

The variation of ϖ with Z for each of the L subshells would be expected to be approximately of the type

$$\varpi_L = (1 + a_L Z^{-4})^{-1}, \tag{3.15}$$

just as for the K fluorescence yield. Departures from this law will be caused by incomplete filling of the outer electron shells for light elements, but such effects should not be important for larger values of Z. If, however, radiationless transitions of the Coster-Kronig type are important in transferring ionization from the L_I shell, say, to the L_{III} shell, there may be serious departures from the relation (3.15) unless allowance has been made for this in the calculation of the fluorescence yield.

Fig. 10 shows an attempt to fit the data on L_{III} fluorescence yield to a relation of type (3.15) for the range of atomic numbers from $Z = 73$ to 92. The agreement is seen to be fair in the case chosen for $a_{L_{III}} = 1 \cdot 02 \times 10^8$.

Similarly, fig. 11 shows the results of Lay's measurements fitted to a relation of this type for $Z = 40$ to 92. The value of a_L corresponding to the curve shown is $a_L = 6 \cdot 4 \times 10^7$. The fit is only fair. However, Lay's results refer to the total fluorescence yield in the L shell. As pointed out above, the relative ionization probabilities of the three L subshells and therefore the total fluorescence yield

TABLE VII. *L-series fluorescence yield*

Element	L_III yield			L_II yield		L_I yield		Total yield				
								Fluorescent excitation			Excitation by γ-ray internal conversion	
	Küstner and Arends (1935)	Stephenson (1937)	Kinsey (1948 a)	Küstner and Arends (1935)	Kinsey (1948 a)	Küstner and Arends (1935)	Kinsey (1948 a)	Lay (1934)	Auger (1925)	Bower (1936)	Kinsey (width estimate) (1948)	Kinsey (counter measurement) (1948)
Kr (36)	—	—	—	—	—	—	—	—	—	—	—	—
Zr (40)	—	—	—	—	—	—	—	0·057	0·13	0·075	—	—
Mo (42)	—	—	—	—	—	—	—	0·067	—	—	—	—
Ag (47)	—	—	—	—	—	—	—	0·100	—	—	—	—
Sb (51)	—	—	—	—	—	—	—	0·119	0·25	0·21	—	—
Te (52)	—	—	—	—	—	—	—	0·122	—	—	—	—
Xe (54)	—	—	—	—	—	—	—	—	—	—	—	—
Ba (56)	—	—	—	—	—	—	—	0·148	—	—	—	—
La (57)	—	—	—	—	—	—	—	0·158	—	—	—	—
Ce (58)	—	—	—	—	—	—	—	0·163	—	—	—	—
Pr (59)	—	—	—	—	—	—	—	0·167	—	—	—	—
Nd (60)	—	—	—	—	—	—	—	0·170	—	—	—	—
Sm (62)	—	—	—	—	—	—	—	0·188	—	—	—	—
Gd (64)	—	—	—	—	—	—	—	0·198	—	—	—	—
Er (68)	—	—	—	—	—	—	—	0·228	—	—	—	—
Hf (72)	0·191	—	0·18	0·326	0·31	0·284	0·17	0·260	—	—	—	—
Ta (73)	0·207	—	—	0·311	—	0·305	—	—	—	—	0·18	—
W (74)	—	—	0·20	—	0·33	—	0·11	0·298	—	—	—	—
Re (75)	—	—	—	—	—	—	—	—	—	—	0·21	—
Os (76)	—	—	—	—	—	—	—	0·348	—	—	0·24	—
Ir (77)	0·244	—	0·23	0·281	0·37	0·370	0·10	—	—	—	—	—
Pt (78)	0·262	—	—	0·274	—	0·392	—	0·348	—	—	0·27	—
Au (79)	0·276	—	0·25	0·272	0·39	0·410	0·09	0·365	—	—	0·29	—
Tl (81)	—	—	0·27	—	0·43	—	0·09	—	—	—	—	—
Pb (82)	0·337	0·32	0·30	0·264	0·46	0·475	—	0·398	—	—	0·32	—
Bi (83)	0·367	—	0·39	0·255	0·56	0·487	—	0·402	—	—	0·40	0·36
Th (90)	—	0·42	0·41	—	0·59	—	—	0·45	—	—	0·42	0·41
U (92)	—	0·44	—	—	—	—	—	—	—	—	—	—

56

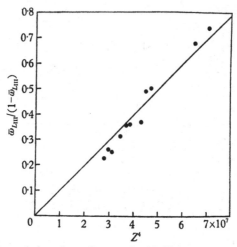

Fig. 10. The variation of $\varpi_{L_{III}}/(1-\varpi_{L_{III}})$ with Z^4 for Z ranging from 73 to 92. The experimental values used are the averaged results of Stephenson, Küstner and Arends, and Kinsey.

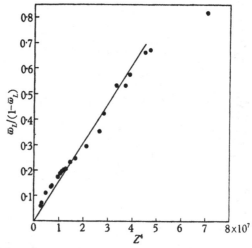

Fig. 11. The variation of $\varpi_L/(1-\varpi_L)$ with Z^4 for Z ranging from 40 to 92. The experimental points are those of Lay for the total L-shell fluorescence yield. The L-ionization was produced by fluorescent excitation.

depends on the frequency of the exciting radiation, and this may influence the nature of the fit of these results to the expression (3.15).

Pincherle's (1935 a) non-relativistic calculations of fluorescence yields included the case of the L shell. For gold his calculated yields for the L_I, L_{II} and L_{III} subshells are respectively 0·154, 0·29 and 0·38. It is clear from Table VII that agreement with the experimental values is not very good.

Further investigations of the L-series fluorescence yields are clearly desirable.

(iii) *The M-shell fluorescence yield*

Very little work has been done on the study of the fluorescence yields for the M shell. Lay (1934), however, obtained a value of 0·06 for the total fluorescence yield of the M shell of uranium.

3.8. The magnetic spectra of the Auger electrons

After Auger had interpreted paired tracks in the cloud chamber in terms of radiationless transitions, it was natural that an attempt should have been made to look for evidence of Auger electrons among the photoelectrons ejected by the incidence of X and other radiation on a solid target. Thus de Broglie and Thibaud (1925) were able to interpret some of the groups of electrons observed in de Broglie's earlier studies of electrons ejected by cathode rays as due to the Auger effect. In this category, the most extensive studies of the magnetic spectra of the Auger electrons ejected when X-rays impinge on various solid materials have been made by Robinson and his collaborators.*

The experimental set-up used in these experiments is shown in fig. 12. H_1 and H_2 (in fig. 12 (a)) are a pair of Helmholtz coils, and K, T are respectively the cathode and target of an X-ray tube of the Muller (gas) type. C is a coil to compensate for the effect of the magnetic field due to the Helmholtz coils on the path of the electrons incident on T. X-rays from T, after passing through an appropriate filter F, pass through the window W into the evacuated chamber B, where they impinge on the disk T′ of the material

* Robinson (1923), Robinson and Cassie (1928), Robinson and Young (1930), Mayo and Robinson (1939).

being studied (see fig. 12 (*b*)). Photoelectrons and Auger electrons ejected from T' describe circular orbits in the magnetic field. After passing through the slit S, electrons of a given energy are brought to a line focus, say at L, on the photographic plate P. It is clear

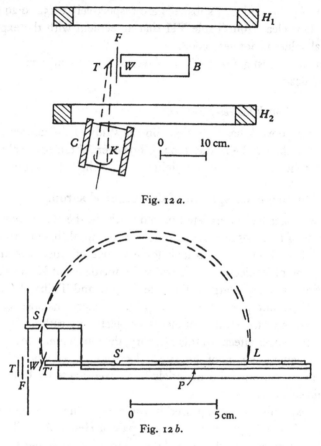

Fig. 12 *a*.

Fig. 12 *b*.

Fig. 12. Robinson's apparatus for studying the magnetic spectra of Auger electrons.

from the diagram that the angular range of the electrons focused at L is determined by the size of T' and its distance from the slit S. S' is a second slit, and by shining a light through it a reference mark can be made on the photographic plate. Then since the positions of S and S' are fixed, the radius of curvature of the electron beam that makes the trace at L is fully determined by the length $S'L$. With an arrangement such as this the targets T' may be changed rapidly without involving difficulties of realinement and adjustment of slits. The targets T' used by Robinson were as thin as convenient, consisting of leaf in some cases and thin layers of suitable compounds in others. No attempt was made, however, to employ truly 'thin' targets in the sense that an inappreciable number of photoelectrons would suffer energy loss prior to escaping. Thus the traces observed on the photographic plate were not lines but bands corresponding to electrons of a range of energy. Estimates of the intensities of the bands on the photographic plate were made visually by comparing them with a set of standards. In this way numbers representing the intensities were assigned. Although it is implied that the numbers assigned to the various bands were roughly proportional to their true intensities there was probably a good deal of uncertainty about this.

Measurements of the magnetic spectrum of Auger electrons have also been carried out for germanium by Ference (1937), using a magnetic spectrograph somewhat similar to that employed by Robinson, but with a Geiger counter for detection. The electrons entered the counter through a celluloid window about 10^{-6} cm. thick. The germanium was deposited as a very thin, barely visible, film on a piece of cellophane, and was illuminated by X-radiation from a silver target, filtered by means of a sheet of palladium. The experimental arrangement used by Ference, employing a Geiger-counter detector and a very thin target source of Auger electrons, should be capable of giving much better measurements of the relative intensities of the various Auger electron 'lines' than were obtained by Robinson and his collaborators. However, in the form used by Ference, the resolving power of the apparatus was inferior to that of Robinson, and Auger lines arising from the interactions of the different groups of L electrons were not resolved.

In Robinson's experiments, no great difficulty was encountered in separating the traces due to Auger electrons and photoelectrons 'externally' ejected by K-series radiation of the same element. That these traces are separate arises from the fact that the energy of the 'external' photoelectrons ejected from a certain level is determined by the quantum energy of the exciting radiation and the energy of that level for a normal atom, while the energy of an Auger electron is determined by the exciting energy and the energy of the corresponding level for an atom already ionized in an inner shell.

For example, Table VIII gives results obtained for the photo-electrons ejected from copper L shells by copper K-radiation and for the Auger electrons from copper ionized in the K shell by molybdenum K-series radiation. These Auger electrons may be thought of as arising from the internal, as opposed to the external, conversion of copper K_{α}-radiation. The ionization energy for the L_{I} level of copper appears from the table as 1·100 keV. for the normal atom and 1·174 keV. for an atom already once ionized in a K shell.

TABLE VIII

Incident radiation	Intensity (relative)	rH	Electron energy (keV.)	Origin of electron trace
MoK (without filter)	2–3	279·5	6·856	$\mathrm{Cu}K_{\alpha_1}$ — 1·174 keV. $(L_{\mathrm{I}}')^*$
	6	283·0	7·023	$\mathrm{Cu}K_{\alpha_1}$ — 1·007 keV. $(L_{\mathrm{II,III}}')^*$
	3–4	300·9	7·934	$\begin{cases} \mathrm{Cu}K_{\alpha_1} - 0\cdot096 \text{ keV. } (M')^* \\ \mathrm{Cu}K_{\beta_1} - 0\cdot951 \text{ keV. } (L_{\mathrm{II,III}}')^* \end{cases}$
CuK (with nickel filter)	6	281·0	6·930	$\mathrm{Cu}K_{\alpha_1}$ — 1·100 keV. $(L_{\mathrm{I}})\dagger$
	4–5	284·3	7·092	$\mathrm{Cu}K_{\alpha_1}$ — 0·938 keV. $(L_{\mathrm{II,III}})\dagger$
	4	301·1	7·945	$\mathrm{Cu}K_{\alpha_1}$ — 0·085 keV. $(M)\dagger$

* Auger.   † Photoelectron.

In this table H oersteds is the magnetic field and r cm. the radius of curvature of the electron beams. The corresponding electron energies in keV. are given in column 4. The probable origins of the various lines are given in column 5. In this column the quantum energy of the $\mathrm{Cu}K_{\alpha_1}$ and $\mathrm{Cu}K_{\beta_1}$ radiations have been taken as 8·030 keV. and of 8·885 keV. respectively. It is interesting to note (column 2 of Table VIII) that not only do Auger electrons and photoelectrons corresponding to similar transitions differ in energy, but they also differ in their relative emission probabilities (intensities).

Robinson's experiments further indicate that, although the relative intensities of the photoelectrons ejected by a given incident radiation from corresponding inner shells of different atoms do not remain constant for elements of widely different atomic number, the relative intensities of the Auger electrons ejected as a result of corresponding transitions in elements of different atomic number are approximately constant.

An interesting feature of the work is that evidence was obtained of Auger transitions of the type $K \to L_I L_I$. If such Auger electrons were to be considered as arising in an internal conversion process, the radiation converted would have to be that corresponding to the forbidden transition $K \to L_I$.

Lines corresponding to Auger electrons have also been observed in the magnetic spectra of electrons from radioactive sources. The mechanism of the emission of these Auger electrons has already been considered. This method of studying the relative intensity of the Auger electrons produced in different transitions has the advantage over Robinson's method that the source may be made extremely thin so that very sharp lines may be obtained.

Ellis (1933 a) studied the Auger electrons ejected after K conversion in an element of atomic number 83, an isotope of bismuth. To identify the element from which the Auger electrons came Ellis studied the magnetic spectra of electrons from Th(B + C) and from Ra(B + C), using photographic detection. Very sharp lines were obtained in these cases because the sources consisted practically of monolayers. The two sources had a certain number of lines in common. From the Ra(B + C) source the Auger electrons observed could only have come from Ra C (83) or Ra C' (84), while from the Th(B + C) source they could have come from Th C (83), Th C″ (81) or Pb (82). Ellis concluded that the Auger lines in common could only have originated from $Z = 83$.

Table IX shows the relative intensities of the lines which he observed.

TABLE IX

Transition	Relative intensity	Transition	Relative intensity
$K \to L_{III} M_{III}$	0·7	$K \to L_{III} L_{III}$	1·5
$K \to L_I N_I$	1·0	$K \to L_{II} L_{III}$	2·8
$K \to L_{II} M_{III}$	0·6	$K \to L_I L_{III}$	1·6
$K \to L_I M_{III}$	0·7	$K \to L_{II} L_{II}$	<0·2
$K \to L_I M_{II}$	0·7	$K \to L_I L_{II}$	2·2
		$K \to L_I L_I$	1·2

Arnoult (1939) and Flammersfeld (1939) have also observed these lines, but Arnoult* differs from Ellis in his assignment of the transitions corresponding to the various lines. Once again the line corresponding to the transition $K \to L_\mathrm{I} L_\mathrm{I}$ was obtained with appreciable intensity.

Steffen, Huber and Humbel (1949) have studied the magnetic spectra of Auger electrons ejected from ^{194}Pt and ^{196}Pt after K capture in ^{194}Au and ^{196}Au (see §3.6). When artificially produced radioactive isotopes are being studied thicker sources have in general to be employed, so that the lines are not so sharp, and it was not possible in this case to separate lines corresponding to different subshells. Table X shows the partial internal conversion coefficients† α for the Pt K series radiation corresponding to a number of Auger transitions.

TABLE X

Transition		α (^{194}Pt)	α (^{196}Pt)
$K \to LL$	(α_{LL})	0·026 ± 0·005	0·028 ± 0·006
$K \to LM$	(α_{LM})	0·018 ± 0·006	0·017 ± 0·006
$K \to LN$	(α_{LN})	0·0065 ± 0·003	0·006 ± 0·003
$K \to MN$	(α_{MN})	0·0065 ± 0·003	0·007 ± 0·003

Table XI shows the relative intensities of the various groups of Auger electrons ejected as a result of interaction between L electrons in K-ionized atoms. In this table the observed relative intensities measured by Robinson and Young (1930) for copper, and by Ellis (1933) for thorium C, are compared with non-relativistic theoretical values, which vary very little with atomic number, and with the relativistic theoretical values calculated for gold.

In view of the approximate nature of the experimental relative intensity estimates for copper the agreement with the non-relativistic theory in this case is very satisfactory. However, the marked change in the relative intensity predicted for heavy elements when relativistic effects are taken into account is not in agreement with the measurements for Th C. Indeed, the relative intensity of the

* Arnoult's interpretation cannot be correct because he assigns the transitions $K \to L_\mathrm{III} L_\mathrm{II}$ and $K \to L_\mathrm{II} L_\mathrm{III}$ to different lines.

† Thus α_{LL} is the ratio of the number of electrons ejected in the process $K \to LL$ to the number of atoms ionized in the K shell.

two Auger groups corresponding to the transitions $K \to L_I L_I$, $K \to L_I L_{II,III}$ in this case is in much better agreement with the non-relativistic calculations.

Table XII gives a comparison between measured relative intensities of the Auger groups arising from the interaction of various L and M electrons and those calculated using a non-relativistic theory (Pincherle, 1935 a).

TABLE XI

Transition	Observed relative intensities		Calculated relative intensities	
	(1) Cu (29)	(2) ThC (83)	(1) Non-relativistic theory	(2) Relativistic theory for Au(79)
$K \to L_I L_I$	1	1	1	1
$K \to L_I L_{II,III}$	5	3·2	3·4	10·8
$K \to L_{II,III} L_{II,III}$	8	3·8	6·7	—

TABLE XII

Transition	Measured relative intensity						Calculated relative intensity (non-relativistic)
	Cu* (29)	Ge† (32)	Sr* (38)	Zr* (40)	Mo* (42)	Pt‡ (78)	
$K \to LL$	1	1	1	1	1	1	1
$K \to L_I M_{I,II,III}$ } $K \to L_{II,III} M$ }	0·6	0·31	0·3	0·2 0·4	0·15 0·3	0·65	0·10 0·48
$K - MM$	—	—	—	0·1	—	—	0·075

* Robinson and Young (1930).   † Ference (1937).
‡ Steffen, Huber and Humbel (1949).

The agreement between the observed and predicted relative intensities is qualitatively satisfactory, but more accurate experimental data is clearly needed. In the case of platinum the agreement may be misleading, since, as previously mentioned, relativistic effects should be appreciable for an element as heavy as this.

CHAPTER IV

THE AUGER EFFECT AND X-RAY SPECTRA

The possibility of radiationless reorganization of an atom ionized in an inner shell has three important consequences in X-ray spectra. It influences the breadths of X-ray emission lines and absorption edges, and also the intensities of X-ray emission lines, and it is one cause of the appearance of the so-called satellite lines arising from transitions in atoms doubly ionized in an inner shell.

4.1. The Auger effect and the breadths of X-ray emission lines and absorption edges

Let p_a^R, p_a^A be respectively the total transition probabilities per unit time for radiative and Auger transitions in an atom initially in a state of inner-shell ionization. The mean lifetime τ_a of the state a is given by $\tau_a = (p_a^R + p_a^A)^{-1}$. The uncertainty principle would then lead to an expected uncertainty Γ_a in the measurement of the energy of the state a, related to τ_a by $\tau_a \Gamma_a \doteq \hbar$, so that the state a would not have a sharply determined energy but would be spread over a range of energy of order of magnitude

$$\Gamma_a = \hbar(p_a^R + p_a^A). \qquad (4.1)$$

Weisskopf and Wigner (1930) have calculated the distribution of probability density in the state a using quantum radiation theory. $P_a(E)\,dE$, the chance that an atom in the state a should have an energy between E and $E + dE$, is given by

$$P_a(E)\,dE = \Gamma_a\,dE / 2\pi\{(E_a - E)^2 + (\tfrac{1}{2}\Gamma_a)^2\}, \qquad (4.2)$$

where E_a is the most probable energy of the state.

Expression (4.2) is of the form of the response curve for a damped harmonic oscillator. The quantity Γ_a is such that when $E = E_a \pm \tfrac{1}{2}\Gamma_a$, $P_a(E)$ has half its maximum value. Γ_a is therefore known as the 'width' of the state.

If an atomic system passes from the state a to a state of lower energy b with the emission of a quantum of radiation of approximate frequency $\nu_{ab} = (E_a - E_b)/h$, Weisskopf and Wigner showed further that the shape of the spectral line emitted has a form similar to (4.2) but with width equal to the sum of the widths of the states, viz. $\Gamma_a+\Gamma_b$. If $I_{ab}(\nu)\,d\nu$ is the energy radiated (per transition) in the frequency range between ν and $\nu+d\nu$, then

$$I_{ab}(\nu) = (\Gamma_a+\Gamma_b)\,\nu/4\pi^2[(\nu_{ab}-\nu)^2+\{(\Gamma_a+\Gamma_b)/2h\}^2]. \quad (4.3)$$

Measurement of the shape of an X-ray absorption edge enables a direct determination to be made of the width of the state of inner-shell ionization to which the absorption edge corresponds. More usually, however, level widths are estimated from the measured breadths of spectral lines.

It is clear from (4.1) that the total width of the state a is made up of two partial widths, viz. $\Gamma_a^R = \hbar p_a^R$, known as the radiation width, and $\Gamma_a^A = \hbar p_a^A$, the Auger width. The fluorescence yield,

$$\varpi_a = p_a^R/(p_a^R+p_a^A),$$

can thus be calculated, if Γ_a^R and Γ_a^A are known, from the relation

$$\varpi_a = \Gamma_a^R/(\Gamma_a^R+\Gamma_a^A). \quad (4.4)$$

The use of (4.4) to determine fluorescence yields has been described in Chapter III (§ 3.7).

The Auger effect accordingly causes an increase in the width of a state above that to be expected from the intensity of radiation emitted in transitions from the state to one of lower energy.

4.2. The Auger effect and X-ray line intensities

Let n_a be the number of atoms per unit time ionized in an inner shell a. When equilibrium has been reached this will also equal the rate at which atoms leave the state a by all possible transitions, with or without radiation, to states of equal or lower energy. Let p_{ab}^R be the transition rate for a radiative transition to the state b. Then corresponding to p_{ab}^R there exists a partial width given by $\Gamma_{ab}^R = \hbar p_{ab}^R$.

The total number, n_{ab}, of radiation quanta corresponding to the transition $a\to b$ per unit time is

$$n_{ab} = n_a\Gamma_{ab}^R/(\Gamma_a^R+\Gamma_a^A). \quad (4.5)$$

From this expression it is clear that the Auger transition probability influences the absolute intensity of X-ray lines produced by an ionizing agent of given strength.

For a second line originating in a different initial state r an expression similar to (4.5) applies, so that the relative intensity of two lines of frequencies ν_{ab}, ν_{rs} corresponding to transitions $a \to b$, $r \to s$ respectively, is given by

$$\frac{I_{ab}}{I_{rs}} = \frac{\nu_{ab} n_a \Gamma_{ab}^R (\Gamma_r^R + \Gamma_r^A)}{\nu_{rs} n_r \Gamma_{pq}^R (\Gamma_a^R + \Gamma_a^A)}, \qquad (4.6)$$

and the Auger effect may have an important influence on the relative intensity of the two lines.

For two lines originating in the same initial level a, however, the relative intensity is
$$I_{ab}/I_{ac} = \nu_{ab} \Gamma_{ab}^R / \nu_{ac} \Gamma_{ac}^R, \qquad (4.7)$$

and is thus independent of the rate of radiationless transitions from the initial state.

4.3. The Auger effect and X-ray satellite lines

X-ray satellite lines arise in transitions in atoms multiply-ionized in inner shells. For example, the normal K-series lines, $K_{\alpha_{1,2}}$ are produced in transitions $K \to L$. The satellite lines K_{α_3}, K_{α_4} are produced in the transitions $KL \to L^2$ and $K^2 \to KL$, respectively.

Atoms doubly ionized in inner shells may be produced directly by electron impact and for K series satellites observed under electron excitation the calculations of R. D. Richtmyer (1936) indicate that the intensity of the satellite relative to that of the parent line and the variation of this relative intensity with atomic number can be explained satisfactorily in terms of direct double ionization by electron impact.

After an Auger transition an atom is left in a state of double inner-shell ionization. Further reorganization of the atom would then be expected to give rise to the emission of $L, M, N, \ldots$ series satellite lines. Let n_a, n_b be respectively the numbers of atoms per unit time ionized directly in the inner shells a and b. We suppose the shell a to have the higher ionization energy. In the subsequent reorganization of the atom the vacancy in this shell may be filled by the transition of an electron from the shell b, either with the emission of radiation or by means of an Auger transition in which an electron in the shell c is removed from the atom. In the former case the atom is left singly ionized in the b shell, while in the latter case it is left doubly ionized in the b and c shells. Consider the X-ray line radiated when the vacancy in the shell b is filled by the transition of an electron from the shell r, of lower ionization energy than b. When a similar transition occurs in the doubly-ionized atom a satellite line of slightly different wave-length is radiated and the intensity $I_{bc,rc}$ of the satellite relative to that, $I_{b,r}$, of the parent line can be calculated.

We denote by Γ_a^R, Γ_a^A the radiation and Auger widths for the state of ionization in the inner shell a, while Γ_b^R, Γ_b^A and Γ_{bc}^R, Γ_{bc}^A are the similar widths for the state of ionization in the shell b and of double ionization in the inner shells b and c respectively. $\Gamma_{a,bc}^A$ is the partial width corresponding to an Auger transition from the initial state to the final state bc of double inner-shell ionization, while $\Gamma_{b,r}^R$, $\Gamma_{bc,rc}^R$ are the partial radiation widths corresponding to the transition $b \to r$ in the singly- and doubly-ionized cases respectively. Then

$$\frac{I_{bc,rc}}{I_{b,r}} = \frac{\nu_{bc,rc}\, n_a \Gamma_{a,bc}^A \Gamma_{bc,rc}^R (\Gamma_b^R + \Gamma_b^A)}{\nu_{b,r}\{n_b(\Gamma_a^R + \Gamma_a^A) + n_a \Gamma_{a,b}^R\}\, \Gamma_{b,r}^R (\Gamma_{bc}^R + \Gamma_{bc}^A)}, \quad (4.8)^*$$

in which $\nu_{bc,rc}$, $\nu_{b,r}$ are respectively the frequencies of the parent and satellite lines. Usually, of course, these frequencies are nearly equal.

If the initial inner-shell ionization is produced by electron impact, and if a and b refer to states of ionization in different inner shells, the ratio n_a/n_b will generally be very small and the contributions of such states to the satellite intensity will also be small.

If, however, a and b refer to states of ionization in different subshells of the same main shell (L_I and L_III for example), n_a/n_b need not be small, and the contribution to the satellite intensity may be comparable to the intensity of the parent line. Satellite lines pro-

---

* If there is more than one state a of higher ionization energy from which the state b may be populated by radiative or Auger transitions, equation (4.8) has to be modified. In this case

$$\frac{I_{bc,rc}}{I_{b,r}} = \frac{\nu_{bc,rc}\, \Gamma_{bc,rc}^R (\Gamma_b^R + \Gamma_b^A)\, \{\sum_a n_a \Gamma_{a,bc}^A/(\Gamma_a^R + \Gamma_a^A)\}}{\nu_{b,r}\, \Gamma_{br}^R (\Gamma_{bc}^R + \Gamma_{bc}^A)\, \{n_b + \sum_a n_a \Gamma_{a,b}^R/(\Gamma_a^R + \Gamma_a^A)\}},$$

in which the summations have to be taken over all states of inner-shell ionization a from which a radiative or Auger transition to a state of inner-shell ionization b is energetically possible.

duced in this way have such an important significance for the interpretation of X-ray spectra that we discuss them in some detail in this chapter.

If the initial state of inner-shell ionization is produced by fluorescence, K capture, or γ-ray internal conversion, n_a/n_b will not generally be small even if a and b refer to states of ionization in different shells. But most studies of X-ray satellites have been carried out under conditions in which the initial inner-shell ionization was produced by electron impact.

4.4. Radiationless transitions of the Coster-Kronig type

The importance of a certain class of radiationless transition in the interpretation of X-ray spectra was first pointed out by Coster and Kronig (1935). An example of a transition of this type is $L_I \rightarrow L_{III} M_{IV,V}$, which may be regarded as arising from internal conversion in the $M_{IV,V}$ shells of radiation emitted in the transition $L_I \rightarrow L_{III}$. Through an understanding of the conditions under which transitions of this type were important Coster and Kronig were able to clear up a number of puzzling anomalies in L-series spectra.

The relative intensities of L-series lines had proved difficult to interpret. For instance, lines such as $L\beta_3$, $L\beta_4$ arising from an L_I initial state of ionization were abnormally weak for many elements and for light elements could not be detected at all, in spite of the fact that other lines originating in the L_{II} and L_{III} levels appeared with considerable intensity.[*]

Further evidence of an anomaly came from a study of the oscillator strengths, f for $2s$ and $2p$ electrons of Ag, Au and Pt, based on measurements of the L absorption spectra of these elements.[†] According to the f-sum rule for oscillator strengths,[‡]

---

[*] O'Bryan and Skinner (1934).

[†] Coster and Kronig, loc. cit.

[‡] The concept of oscillator strength came from the classical picture of the interaction between radiation and an atomic system. Each critical frequency of an atomic system was interpreted as arising from an atomic oscillator of that frequency. Since the number of critical frequencies is greatly in excess of the number of electrons, each electron had to be associated with a number of oscillators. If ϵ_n is the effective charge of one of these oscillators, $\epsilon_n^2 = e^2 f_n$, where f_n is the corresponding oscillator strength. If ψ_0, ψ_n are the wave functions for the initial and final states for a transition, and if ν_n is the corresponding

the sum of all the f's for all possible transitions for any one electron should be 1. In estimating this sum for an electron in a many-electron atom, allowance has to be made for the inaccessibility of certain occupied states owing to the application of the Pauli principle. Even though inaccessible, these states have to be taken as contributing to the f-sum. Thus the sum of all the oscillator strengths for transitions from a given state, a is equal to the number of electrons in that state. The sum of the oscillator strengths for all transitions from the $L_1(2s)$ state should accordingly be 2, and from the $(L_{II} + L_{III})(2p)$ state, 6, and the ratio of the f-sums for the $2s$ and $2p$ states should be 0·33. The actual value of this ratio deduced from the L absorption measurements was nearer 0·25.

Further, emission lines arising from an L_I initial state generally had a width considerably greater than lines arising from L_{II} and L_{III} initial states.

All these observations seemed to point to the existence of some unobserved transition that removed atoms in an L_I state. Several investigators had in fact searched for radiation from the allowed transitions $L_I \rightarrow L_{II, III}$ but had failed to observe it.* Evidently it was too weak to account for the much lower population and rapid depletion of atoms ionized in the L_I shell in comparison with those ionized in the L_{II} and L_{III} shells.

The clue to the interpretation of these anomalies came when Coster and Kronig connected them with the behaviour of the L-series satellites. These satellites show several important differences from K-series satellites. They are often much more intense relative to the parent line than the latter, and this relative intensity varies in a somewhat irregular way with atomic number. Further, L-series satellites can be produced by fluorescent excitation, whereas K-series satellites have only been observed in X-ray spectra excited by electron impact. Coster and Kronig suggested

absorption frequency, then the quantum theoretical expression is

$$f_n = 4\pi m \nu_n |\mathbf{M}_{0n}|^2 / e^2 \hbar,$$

where $\mathbf{M}_{0n}$ is the electric dipole moment corresponding to the transition and is given by

$$\mathbf{M}_{0n} = e \int \psi_n^* \mathbf{r} \psi_0 dr.$$

The proof that $\Sigma f_n = 1$ for all transitions for a given electron is given in Mott and Sneddon (1948, p. 169).

* See, however, Tomboulian (1948).

that they arose from transitions in atoms ionized in the L and M shells, and that such states of ionization were produced, in atoms initially ionized in the L_I shell by an Auger transition of the type $L_I \to L_{III} M_{IV,V}$. Such a process is not energetically possible for atoms in the range of atomic number $Z = 50$ to $Z = 74$. It is just in this range that the satellites of $L\alpha_1$ and $L\beta_2$ are too weak to be observed, while for atomic numbers outside this range their intensity varies from a few to about 50 % of the parent-line intensity.

Since the probability of initial ionization in the L_I shell is of the same order as that for L_{II} and L_{III} shell ionization, an Auger effect of the Coster-Kronig type, if it can occur with appreciable probability, will disturb seriously the intensity relations in L-series spectra.

4.5. The transition rate for Coster-Kronig transitions

It is possible to see qualitatively the circumstances under which the Coster-Kronig transition rate is likely to be large. Using expression (2.1) of Chapter II for the Auger transition rate we take for the wave functions

$$\chi_i(\mathbf{r}_1) = f_{2s}(r_1)/r_1,$$
$$\chi_f(\mathbf{r}_1) = f_{2p}(r_1) P_1(\cos\theta_1)/r_1,$$
$$\psi_i(\mathbf{r}_2) = f_{3d}(r_2) P_2(\cos\theta_2)/r_2,$$
$$\psi_f(\mathbf{r}_2) = \sum_l (2l+1) f_l(kr_2) P_l(\cos\vartheta)/kr_2,$$

where $f_{2s}(r)/r, f_{2p}(r)/r$ and $f_{3d}(r)/r$ are the radial parts of the wave functions of the electron in the $2s$, $2p$ and $3d$ states, respectively, and

$$f_l(kr) \sim \sin(kr - \tfrac{1}{2}l\pi + \delta_l).$$

For simplicity we consider only the case where the magnetic quantum number, $m = 0$. ϑ is the angle between the radius vector $\mathbf{r}_2$ and the direction of ejection which we suppose to make an angle α with the polar axis. The energy of the ejected electron E is given by $E = k^2\hbar^2/2m$.

Further,

$$1/|\mathbf{r}_1 - \mathbf{r}_2| = \sum_{n=0}^{\infty} a_n(r_1, r_2) P_n(\cos\Theta),$$

where $\qquad\qquad a_n = r_1^n/r_2^{n+1} \quad \text{for} \quad r_1 < r_2$

and $\qquad\qquad\quad = r_2^n/r_1^{n+1} \quad \text{for} \quad r_2 < r_1,$

and Θ is the angle between the radii vectores to the electrons 1 and 2 respectively.

Expression (2.1) will then give the transition rate into unit solid angle in the direction α and will have to be integrated over all angles of ejection to obtain the total Auger transition rate.

Using the expansions

$$P_l(\cos\vartheta) = P_l(\cos\theta_2)P_l(\cos\alpha)$$
$$+ 2\sum_{m=1}^{l} \{(l-m)!/(l+m)!\} P_l^m(\cos\theta_2)P_l^m(\cos\alpha)\cos m(\phi_2 - \beta),^*$$

and a similar expression for $P_n(\cos\Theta)$, the angular integrations that have to be carried out are

$$\int\int P_n(\cos\Theta)\,P_1(\cos\theta_1)\sin\theta_1\,d\theta_1\,d\phi_1, \tag{4.9}$$

which clearly vanishes unless $n = 1$, and

$$\int\int P_1(\cos\theta_2)\,P_2(\cos\theta_2)\,P_l(\cos\vartheta)\sin\theta_2\,d\theta_2\,d\phi_2, \tag{4.10}$$

which vanishes unless $l = 1$ or 3.

The radial integrals that have to be carried out are thus of the type

$$\int\int a_n(r_1, r_2)f_{2s}(r_1)f_{2p}(r_1)f_{3d}(r_2)f_l(kr_2)\,dr_1\,dr_2, \tag{4.11}$$

with $l = 1$ or 3 and $n = 1$.

Fig. 13 shows the forms of the functions f_{2s}, f_{2p}, f_{3d} calculated from the Hartree field of a typical atom of medium atomic number ($Z \doteq 50$). The functions $f_{2s}(r_1)$ and $f_{2p}(r_1)$ overlap considerably, thus ensuring that the integral over r_1 will be fairly large.

Fig. 13 shows also a typical form of $f_l(kr)$ ($l = 3$). If a zero of f_l falls in the region of r_2 for which $f_{3d}(r_2)$ is large, there will be a great deal of cancellation in the r_2 integration and the Auger transition rate will be small. For very small values of the energy of the ejected electron only the $l = 1$ case will be important and the Auger transition probability will increase with k. As k increases, however, the first zero of $f_1(kr_2)$ occurs for smaller values of r_2 and eventually falls in the region where $f_{3d}(r_2)$ is large. The major contributions to (4.11) then come from $l = 3$. Finally, when k increases still further a great deal of cancellation occurs in the r_2 integration for all values of l, and the Auger transition probability becomes small again. Since the kinetic energy of the ejected electron will vary considerably with Z the probability of transitions of the Coster-Kronig type will show a marked dependence on atomic number.

To get the total Auger transition rate the amplitude for the 'exchange' process $3d \to 2s$, $2p \to \infty$ should also be calculated and combined by use of expression (2.12), with that obtained for the 'direct' process $2p \to 2s$, $3d \to \infty$ discussed above. The radial integrals that have to be carried out in the 'exchange' case are of the type

$$\int\int a_n(r_1, r_2)f_{2s}(r_1)f_{3d}(r_1)f_{2p}(r_2)f_l(kr_2)\,dr_1\,dr_2,$$

---

* The angle β defines the azimuth of the ejected electron.

with $l = 1$ or 3 and $n = 2$. From fig. 13 it is seen that f_{3d} and f_{2s} do not overlap very much, so that the amplitude will be considerably smaller than in the 'direct' case. Destructive interference between $f_l(kr_2)$ and $f_{2p}(r_2)$, however, becomes important at considerably larger values of k than for the 'direct' case, so that the 'exchange' amplitude should vary much less rapidly with change of the kinetic energy of the ejected electron, and therefore with atomic number Z.

The variation of the total Auger transition rate for elements near the atomic number for which the process is just energetically possible will be determined mainly by the variation of the amplitude for the direct transition.

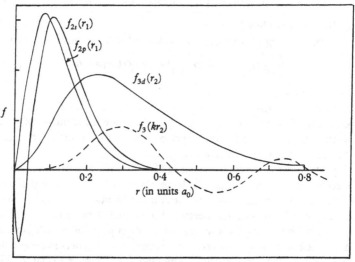

Fig. 13. Form of the radial functions $f_{2s}(r_1)$, $f_{2p}(r_1)$, $f_{3d}(r_2)$, $f_3(kr)$ important in the calculation of the Coster-Kronig transition probability for the transition $(L_I \rightarrow L_{III} M_{IV, V})$.

4.6. Possible types of Coster-Kronig transitions

If E_{a_1}, E_{a_2} are the ionization energies of the two sublevels concerned in a Coster-Kronig transition and E_b' that of the outer level from which the electron is ejected, the kinetic energy of this electron is

$$E_{a_1} - E_{a_2} - E_b',$$

and the process is energetically possible only if

$$E_{a_1} - E_{a_2} > E_b'. \qquad (4.12)$$

Fig. 14 shows the variation of $E_{a_1} - E_{a_2}$ and E_b with Z for a number of the most important Coster-Kronig transitions. In par-

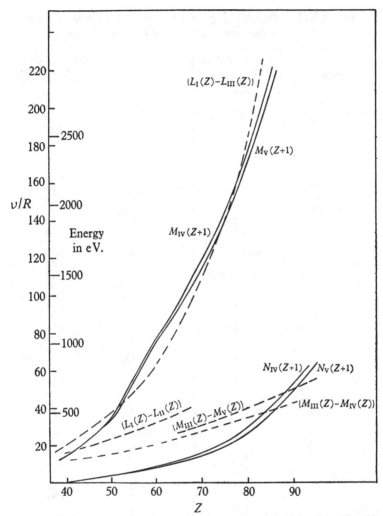

Fig. 14. Variation with Z of the ionization energies of the M_{IV}, M_V, N_{IV}, N_V shells, and the differences in ionization energies L_I–L_{II}, L_I–L_{III}, M_{III}–M_{IV}, M_{III}–M_V, to illustrate the range of atomic numbers over which various Coster-Kronig transitions can occur.

ticular, it shows that the transitions $L_I \rightarrow L_{III}M_{IV,V}$ is energetically possible for $Z < 50$ and again for $Z > 73$, but not in the intermediate region. Similarly, the transition $M_{III} \rightarrow M_{IV}N_{IV,V}$ can occur for $Z < 84$.

TABLE XIII. *Possible Auger transitions of the Coster-Kronig type and the elements in which they may occur*

Transition	Values of Z for which transition is energetically possible	Transition	Values of Z for which transition is energetically possible
$L_I \to L_{II} N_I$	< 70	$M_{III} \to M_{IV} M_{IV}$	< 36
N_{II}	< 75	$M_V N_V$	< 91
N_{III}	< 81	N_{IV}	< 89
N_{IV}	< 92	N_{III}	< 77
$L_I \to L_{II} M_I$	< 31	N_{II}	< 70
M_{II}	< 36	N_I	< 57
M_{III}	< 37	M_V	< 37
M_{IV}	< 40	$M_{IV} \to M_V O_{III}$	< 85
M_V	< 40	O_{II}	< 82
$L_I \to L_{III} M_I$	< 31	$N_I \to N_{II} O_{III}$	< 88
M_{II}	< 36	O_{II}	< 85
M_{III}	< 37	O_I	< 79
M_{IV}	< 50 and > 73	N_{VII}	< 80
M_V	< 50 and > 77	N_{VI}	< 80
$L_{II} \to L_{III} M_{IV}$	< 30	N_V	< 53
M_V	< 30 and > 90	N_{IV}	< 53
$M_I \to M_{II} N_{VII}$	< 91	$N_I \to N_{III} N_{VI}$	< 92
N_{VI}	< 91	N_V	< 53
N_V	< 72	N_{IV}	< 53
N_{IV}	< 71	$N_{IV} N_V$	< 84
N_{III}	< 53	N_{IV}	< 80
N_{II}	< 53	$N_V N_V$	< 87
N_I	< 47	$N_{II} \to N_{III} O_{III}$	< 55 and > 66
M_V	< 34	$N_{III} O_{III}$	< 55 and > 66 and < 86
M_{IV}	< 34	N_{VII}	< 80
$M_I \to M_{III} N_V$	< 87	N_{VII}	< 80
N_I	< 50	$N_{IV} N_V$	< 60
$M_{V'}$	< 34	N_{IV}	< 60
M_{IV}	< 34	$N_V N_V$	< 62
$M_{IV} M_V$	< 44	$N_{III} \to N_{IV} O_I$	< 87
M_{IV}	< 43	N_{VII}	< 86
$M_V M_V$	< 45	N_{VI}	< 85
$M_{II} \to M_{III} N_{IV}$	< 48 and > 65	N_V	< 57
N_V	< 48 and > 65	N_{IV}	< 56
$M_{IV} N_I$	< 88	$N_V O_I$	< 89
M_V	< 35	N_{VII}	< 87
M_{IV}	< 35	N_{VI}	< 86
$M_{III} \to M_{IV} N_V$	< 85	N_V	< 57
N_{IV}	< 84	$N_{IV} \to N_V O_{IV, V}$	< 81
N_{III}	< 71	$N_V \to N_{VI} N_{VII}$	< 91
N_{II}	< 65	N_{VI}	< 91
N_I	< 55	$N_{VII} N_{VII}$	< 91
M_V	< 36		

Many other Coster-Kronig transitions are possible for various ranges of Z. Table XIII given by Cooper (1944) lists all the possible processes of this type and the range of Z over which they should occur. This table is based on the term values given by Siegbahn

(1931). In the curves of fig. 14 and the figures of Table XIII the energy taken for the two inner levels was that corresponding to an atomic number Z, while that for the outer level corresponded to an atomic number $Z+1$. This procedure is justified, since owing to the vacancy in the inner shell the outer electron will be moving in a field approximately equivalent to that corresponding to atomic number $Z+1$. The energy of the state of double ionization is not, however, given exactly by this procedure. This causes some uncertainty in the atomic number for which a given Auger transition can just occur. Some idea of the error involved in this procedure can be obtained from Table XIV compiled by Cauchois (1944). From an analysis of satellite line frequencies* Cauchois constructed an energy-level diagram for an atom doubly ionized in inner shells. Table XIV shows $\Delta E_{L_{III}}$, the difference in L_{III} ionization energy between an atom already ionized in the $M_{IV,V}$ shell and a normal atom. In terms of our previous notation

$$\Delta E_{L_{III}} = (E_{L_{III}, M_{IV,V}} - E_{M_{IV,V}}) - E_{L_{III}}$$
$$= (E_{L_{III}} + E'_{M_{IV,V}} - E_{M_{IV,V}}) - E_{L_{III}} = E'_{M_{IV,V}} - E_{M_{IV,V}}$$
$$(4.13)$$

where $E'_{M_{IV,V}}$ is the $M_{IV,V}$ ionization energy of an atom already ionized in the L_{III} shell. In fig. 14 we have written

$$E'_{M_{IV,V}}(Z) = E_{M_{IV,V}}(Z+1). \qquad (4.14)$$

If this were exactly true,

$$\Delta E_{L_{III}} = E_{M_{IV,V}}(Z+1) - E_{M_{IV,V}}(Z). \qquad (4.15)$$

Both $\Delta E_{L_{III}}$ as estimated by Cauchois and $E_{M_{IV,V}}(Z+1) - E_{M_{IV,V}}(Z)$ are given in Table XIV. Their approximate equality shows that our assumption (4.14) is justified.

Table XIV also shows $\Delta E_{M_{IV,V}}$, $\Delta E_{N_{IV,V}}$ and $\Delta E_{O_{IV,V}}$, the differences in $M_{IV,V}$, $N_{IV,V}$ and $O_{IV,V}$ ionization energies respectively between an atom already ionized in the $M_{IV,V}$ shell and a normal atom.†

* Comprehensive tables of all known satellites have been prepared by Cauchois and Hulubei (1947).

† The values of $\Delta E_{L_{III}}$ estimated by Cauchois can be compared with the values deduced from Robinson's measurement of the difference in energy of Auger and photo-electrons corresponding to similar transitions (see § 3.8). Thus for gold, Robinson's measurements give a value of $\Delta E_{L_{III}}$ of about 70 eV. in reasonable agreement with that estimated by Cauchois.

Referring to fig. 14 for the transition $L_I \to L_{III} M_{IV,V}$, it is seen that the kinetic energy E of the ejected electron increases from zero for Z a little above 50 to about 100 eV. for $Z = 41$ and increases rapidly for smaller Z. On the basis of the arguments set out in §4.5 Coster and Kronig estimated that the Auger transition rate for this transition should reach a maximum for an ejected electron energy of about 100 eV. The occurrence of this maximum for such small ejected electron energies implies that each of the possible transitions set out in Table XIII will only occur with appreciable probability for values of Z near the 'cut-off' for that particular transition.

TABLE XIV. *Difference of X-ray energy levels in electron-volts between atoms ionized doubly and singly in an inner shell*

Element	$\Delta E_{L_{III}}$		ΔE_{M_V}		ΔE_{N_V}		ΔE_{O_V}		$E_{M_{IV}}(Z+1)$ $-E_{M_{IV}}(Z)$	$E_{M_V}(Z+1)$ $-E_{M_V}(Z)$
	A	B	A	B	A	B	A	B		
Pt (78)	89	67	52	—	19	12	(0)	(0)	94	89
Au (79)	92	73	53	47	19	17	0·4	0·4	94	89
Tl (81)	98	80	55	50	21	22	1·1	0·8	108	95
Pb (82)	101	81	57	50	23	21	2·0	2·0	99	92
Bi (83)	104	85	60	54	27	(42)	2·5	1·9	—	—
Ra (88)	116	99	65	57	28	33	4·0	3·9	—	—
Th (90)	125	—	70	—	36	—	6·3	—	122	109

The A and B refer to values of the estimated energy difference obtained in two different ways.

4.7. Interpretation of *L*- and *M*-series satellites by the Coster-Kronig theory

The Coster-Kronig theory has proved remarkably successful in accounting for the intensities of *L*- and *M*-series satellites and their variation with atomic number. Referring to equation (4.8), if a and b refer to two levels in the same shell n_a and n_b will be of comparable magnitude if the ionization is produced by electron impact.* Further, if the energy of the ejected electron is of the right order of magnitude, the ratio $\Gamma^A_{a,bc}/(\Gamma^R_a + \Gamma^A_a)$ may be comparable to 1 and the ratio $\Gamma^R_{a,b}/(\Gamma^R_a + \Gamma^A_a)$ may be small. The ratios $\Gamma^R_{bc,rc}/\Gamma^R_{b,r}$ and $(\Gamma^R_{bc} + \Gamma^A_{bc})/(\Gamma^R_b + \Gamma^A_b)$ will also generally be

* See, for example, Burhop (1940).

about 1, so that for some values of Z the intensity of a Coster-Kronig satellite may be comparable with that of the parent line.

Plate III shows some typical L series spectra of heavy elements obtained by Cauchois. The diagram and satellite lines are indicated.

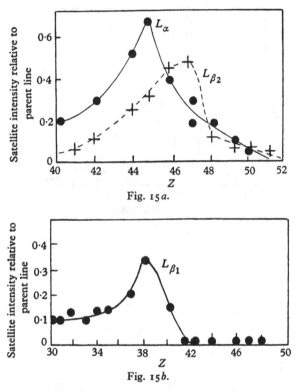

Fig. 15 a.

Fig. 15 b.

Fig. 15. Variation with Z of the intensity of L-series satellites relative to that of the parent line. (a) $L\alpha$ (Hirsch, 1935); $L\beta_2$ (Pearsall, 1934); (b) $L\beta_1$ (Hirsch, 1936).

Fig. 15 shows the variation with Z of the relative satellite to parent intensity for the satellites $L\alpha_1(L_{III} \to M_V)$, $L\beta_1(L_{II} \to M_{IV})$ and $L\beta_2(L_{III} \to N_V)$. In each case the sum of all the satellite intensities is shown as a fraction of the intensity of the parent line. The absolute value of this ratio is a difficult quantity to measure, because the satellites have to be separated from the background due to the parent line, and the measured value of the satellite intensity

will depend on the criteria employed to define this background·
This can be seen from fig. 16, which shows microphotometer traces
of the $L\alpha$ satellites of Ag and Pb. For the Ag $L\alpha_1$ satellites, values of
0·155,[*] 0·33[†] and 0·08[‡] for the relative intensity have been found
by different observers.

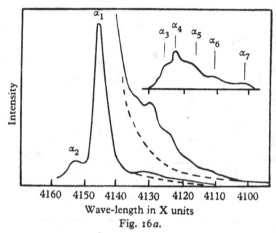

Fig. 16a.

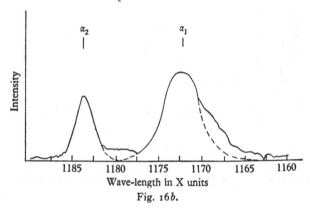

Fig. 16b.

Fig. 16. Microphotometer traces of $L\alpha$ satellites. (a) Ag$L\alpha$ (Richtmyer, 1937);
(b) Pb$L\alpha_1\alpha_2$ (Valadares, 1940). The inset curve in (a) shows the satellite
structure after the background due to the parent line has been subtracted.

However, although different observers may disagree in absolute
value they all agree that the variation of relative satellite intensity
with atomic number has the characteristic form shown in fig. 15.

[*] Hirsch (1935).   [†] Hirsch and F. K. Richtmyer (1933).
[‡] Randall and Parratt (1940).

PLATE III

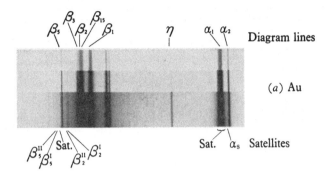

(a) Au

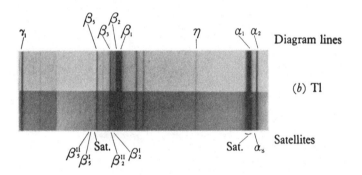

(b) Tl

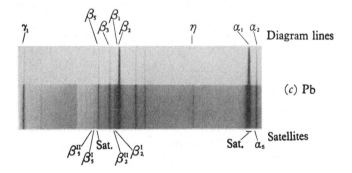

(c) Pb

For explanation see p. xi

In each case the satellite intensity is too small to detect for elements for which the relation (4.12) is not satisfied. Particularly striking is the reappearance of the satellites of $L\alpha_1$, for values of $Z > 73$, after being negligible between $Z = 52$ and 73 as expected from the above considerations.

The satellite intensity increases to a sharp maximum near the critical atomic number for 'cut-off'.* For example, for the $L\alpha_1$ satellites this maximum occurs for $Z = 45$ near where it would be expected from the considerations of §4.5.

A direct test of the Coster-Kronig theory can be made by measuring the excitation potential for appearance of the satellite lines. This excitation potential has been measured for the $L\alpha_1$ satellites of Nb(41) by Coster, Kuipers and Huizinga (1935) and of Au(79) by Valadares and Mendes (1948). These satellites are attributed to the transition $L_{III}M_{IV,V} \to M_V M_{IV,V}$. If they were produced by direct double ionization in the L_{III} and $M_{IV,V}$ shells, their excitation potential should be just the sum of the ionization potentials of those shells. If, however, the initial state is produced by a Coster-Kronig transition from a state of inner ionization in the L_I shell, then excitation potential should be the L_I ionization potential. For Nb(41) the ionization potentials of the L_I and L_{III} shells are respectively 2700 and 2360 V. The ionization potentials of the $M_{IV,V}$ shells are approximately 240 V. If then the satellites are excited by a direct double-ionization process their excitation potential should be approximately 2600 V. The observed excitation potential was 2700 V., in agreement with the Coster-Kronig theory of the origin of these satellites. The similar experiment for gold is more difficult, since in this case the L_I ionization potential is 14·4 kV., while the sum of the L_{III} and M_V ionization potentials is 14·2 kV. However, Valadares and Mendes (1948) interpret their observations as also providing support for the Coster-Kronig theory.

* The only case in which an expected satellite intensity anomaly has been searched for and not found is that for M_β ($M_{IV} \to N_{VI}$) for which the expected 'cut-off' value of Z is 84. The satellites of M_β have an intensity about 2 % of that of the parent line for Z between 77 and 92, and this is of the order of magnitude that might be expected to be produced by direct double ionization. Hirsch (1942) has sought to interpret the non-occurrence of the anomaly in this case as due to the high probability of a competing Auger transition that does not lead to the appearance of one of the known satellites of M_β.

A detailed calculation of the $L\alpha$ and $L\beta_2$ satellite structure and the satellite intensity relative to the parent lines has been carried out by F. K. Richtmyer and Ramberg (1937) for gold. The calculation is very complicated because of the number of satellite components. For the $L\alpha$ satellites corresponding to transitions between the doubly-ionized states $2p_{\frac{3}{2}}3d \rightarrow 3d^2$, assuming j-j coupling, there are 29, and for the $L\beta_2$ satellites, there are 40 components in the resultant super-multiplet.

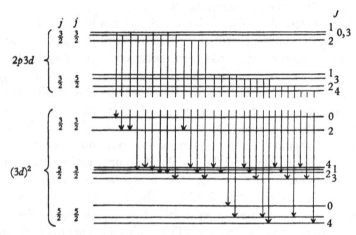

Fig. 17. Possible transitions between the doubly ionized states $2p_{\frac{3}{2}}3d \rightarrow 3d^2$ giving rise to the L_α satellites (F. K. Richtmyer and Ramberg, 1937).

Fig. 17 shows the transitions giving rise to the components of the $L\alpha$ satellites. The main part of the calculation of Richtmyer and Ramberg consisted in the determination of the separations of the various components from the parent line. This was done, assuming the electrons to move in the Thomas-Fermi field for $Tl^{++}(81)$. This field was chosen since for the outer electrons at least it agreed closely with the field of a gold atom doubly ionized in inner shells. The relative intensities of the satellite components were obtained from formulae given by Bartlett (1930) for the relative intensities of super-multiplet lines. The width of each component of the satellite was obtained by adding the widths of the initial and final doubly-ionized states. For the $L\alpha$ satellites the width of the initial state was taken as the sum of the widths of L_{III} and $M_{IV,V}$, while that of the final state was taken to be twice the width of the $M_{IV,V}$

state. A similar procedure was followed for the $L\beta_2$ satellites. This procedure assumes that the absence of the second electron does not affect appreciably the chance per unit time of filling the other vacancy, and this assumption appears justified.

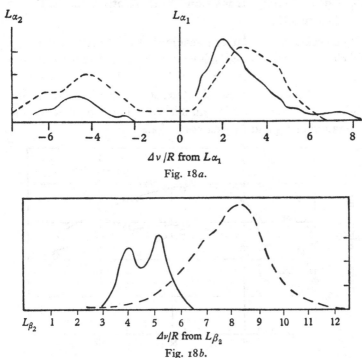

Fig. 18a.

Fig. 18b.

Fig. 18. Comparison of observed and calculated satellite intensity distribution· —— observed; ---- calculated. (a) Pb $L\alpha_1\alpha_3$ (Valadares, 1940); (b) Au $L\beta_2$ Richtmyer *et al.* (1934). The calculations were carried out for Au, but the distribution for Pb should be very similar.

Fig. 18 shows a comparison of the calculated satellite structure with that observed for Pb $L\alpha_1\alpha_2$ (Valadares, 1940) and Au $L\beta_2$ Richtmyer *et al.* (1934). The calculated structure for the satellites of $L\alpha$ agrees surprisingly well with that observed, in view of the inexactness of the assumed atomic field. For $L\beta_2$ the agreement is not quite so good, but in view of the nature of the assumptions involved it is still reasonable.

Similar calculations have been carried out by Pincherle (1942) for the $L\alpha$ and $L\beta$ satellites of elements in the range of atomic number from $Z = 37$ to 56. Once again good qualitative agreement was obtained with the experimental results, but the calculated separations of the components were greater than those observed.

4.8. Coster-Kronig transitions and relative intensities of L-series lines

The occurrence of Auger transitions of the Coster-Kronig type will clearly have a marked influence on L-series line intensities. If such transitions can occur, the total Auger width, in equation (4.6), is increased for atoms in the L_I state, so that the intensity of lines originating in this state is diminished relative to that of lines originating in L_{II} and L_{III} states (see §4.4).

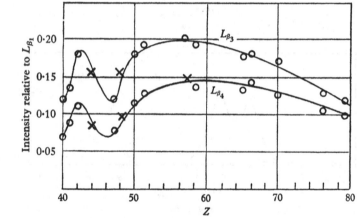

Fig. 19. Variation with atomic number of the intensities of the lines $L\beta_3$ and $L\beta_4$ relative to that of $L\beta_1$. O de Langen (1939, 1940); × Coster and Bril (1942).

Measurements of the intensities of the lines $L\beta_3 (L_I \to M_{III})$ and $L\beta_4 (L_I \to M_{II})$ relative to $L\beta_1 (L_{II} \to M_{IV})$ have been carried out for a wide range of values of Z.* Fig. 19 shows the results of these measurements, which receive a ready interpretation on the basis of the Coster-Kronig theory. For $Z = 40$ the Coster-Kronig transition probability for $L_I \to L_{II}M_{IV,V}$ is about at a maximum (see fig. 15), and the relative intensities of the lines originating in the L_I state is a minimum. For $Z = 42$ this radiationless transition is no longer possible, and consequently the intensity of $L\beta_3$ and $L\beta_4$ relative to

* Coster and de Langen (1936), de Langen (1939, 1940); Coster and Bril (1942), Bril (1947), Cooper (1942).

$L\beta_1$ has risen again. As Z increases further, however, another Coster-Kronig transition, $L_I \to L_{III}M_{IV,V}$, already energetically possible, becomes important, its probability reaching a maximum for $Z = 45$ or 46 (see fig. 15). Accordingly, the curves of fig. 19 fall to a minimum about this point. For $Z = 52$ the transition $L_I \to L_{III}M_{IV,V}$ in its turn is excluded on energetic grounds, and the curves rise again to values about the same as for $Z = 42$. Another rapid (but less marked) fall of relative intensity with increasing Z occurs for $Z > 73$, when the transition $L_I \to L_{III}M_{IV,V}$ is again possible.

Measurements have also been made* of the intensities of $L\beta_3$, $L\beta_4$ relative to $L\beta_1$ and of $L\gamma_2$ ($L_I \to N_{II}$) and $L\gamma_3$ ($L_I \to N_{III}$) relative to $L\gamma_1$ ($L_{II} \to N_{IV}$) for $Z > 70$. In each case the relative intensity falls gradually with increasing Z for $Z > 73$. On the other hand, the intensity of $L\beta_2$ ($L_{III} \to N_{VI}$) relative to $L\beta_1$ ($L_{II} \to M_{IV}$) remains constant in this range, as would be expected.

Using equation (4.6) and the curves of fig. 19 it is possible to estimate the fraction of atoms ionized in the L_I shell which undergo a radiationless transition of the Coster-Kronig type. For silver this fraction comes out to be nearly 0·5.

Several measurements have been made of the spectrum of the γ-rays and X-rays which accompany the radioactive decay of Ra D (Frilley and Tsien, 1945; Curran, Angus and Cockcroft, 1949; Salgueiro and Valadares, 1949). The L-series radiations of Ra E (83) have been identified, and these have been attributed to radiations emitted following the internal conversion of the γ-radiation in one of the L subshells of Ra E. The relative intensities of the $L\alpha_1$, $L\beta_{1,2}$, $L\beta_3$ and $L\beta_4$ lines determined by Salgueiro and Valadares (1949) for Ra E were 100:95:40:45. The corresponding values for excitation by electron impact are 100:73:6·5:5. This difference in relative intensities provides support for the interpretation of the X-ray spectrum as arising from the internal conversion of the γ-radiation of Ra D. For the relative number of vacancies in the L_I and L_{III} shells is 100:232 when the mode of ionization is by electron impact, compared with 100:3·1 when the inner-shell ionization arises from internal conversion in Ra E.

Salgueiro and Valadares used equation (4.6) to calculate the

* Cooper (1942).

expected relative intensities of the L-series lines when excited by internal conversion, and estimated for the relative intensity of the α_1 and β_4 lines the ratio 100:383. This calculation took no account, however, of the possibility of Coster-Kronig transitions between the L_I and L_{III} subshells. When allowance was made for the transfer of ionization from the L_I to the L_{III} shells by such radiationless transitions, the relative intensity of those lines excited by internal conversion was calculated to be 100:35, in fairly good agreement with the observed value of 100:45. This calculation provides further striking evidence of the important role of Coster-Kronig transitions in determining X-ray line intensities in many cases.

4.9. Coster-Kronig transitions and widths of L and M levels

When Coster-Kronig transitions can occur with appreciable probability the width of the L_I level is increased considerably. Fig. 20 shows the structure of the L_I, L_{II} and L_{III} absorption edges as measured by Richtmyer, Barnes and Ramberg (1934) for gold, illustrating this effect. From the shape of these absorption edges it was possible to calculate the width, Γ. Assuming that in the absorption process electrons are ejected into unoccupied levels at the top of the Fermi distribution, and that these levels are uniformly distributed in frequency, the absorption coefficient for radiation of frequency ν is given by

$$\mu(\nu) = C'[\pi - 2\arctan\{2(\nu_0 - \nu)/\gamma\}]/2\pi, \qquad (4.16)$$

where ν_0 is the frequency at the centre of the absorption band, and the width γ of the state responsible for the absorption is now expressed on a frequency scale.*

---

* If the correct Fermi distribution is assumed, the absorption coefficient $\mu(\nu)$ is given by

$$\mu(\nu) = A'[\arctan\{B/((2\nu_i)^{\frac{1}{2}} - A)\} + \arctan\{B/((2\nu_i)^{\frac{1}{2}} + A)\}]$$
$$+ \tfrac{1}{2}B' \log\{(\nu_i + D + (2\nu_i)^{\frac{1}{2}} A)/(\nu_i + D - (2\nu_i)^{\frac{1}{2}} A)\}, \quad (4.17)$$

where $h\nu_i$ is the difference between the Fermi limit and the mean energy of electrons in the metal,

$$A = [D + (\nu_i + \nu - \nu_0)]^{\frac{1}{2}}, \quad B = [D - (\nu_i + \nu - \nu_0)]^{\frac{1}{2}},$$
$$D = [(\nu_i + \nu - \nu_0)^2 + (\Gamma/2)^2]^{\frac{1}{2}}, \quad A' = CA, \quad B' = CB,$$

where C is a constant and

$$0 \leqslant \arctan[B/\{(2\nu_i)^{\frac{1}{2}} \pm A\}] \leqslant \pi.$$

Level widths deduced from this exact formula do not differ appreciably from those calculated from the approximate expression (4.13).

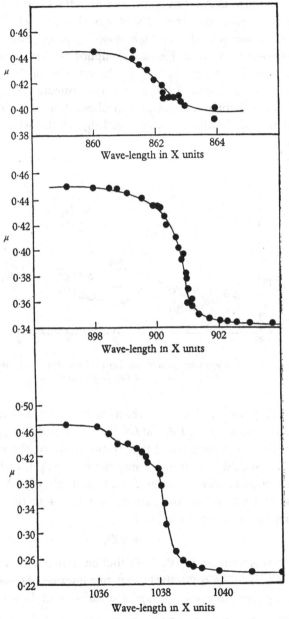

Fig. 20. Structure of the L_I, L_{II} and L_{III} absorption edges of Au·
(Richtmyer, Barnes and Ramberg, 1934).

Most of the available data on the widths of the L levels, however, are derived from measurements of emission-line widths, using X-ray spectrometers of very high resolving power and not from absorption-edge studies. Fig. 21, compiled by Bril (1947) from data of several investigators,* shows the variation of the widths of the lines $L\beta_1$, $L\beta_2$, $L\beta_3$ and $L\beta_4$ with Z. Unfortunately, it is difficult to deduce unambiguous information about L-level widths from data of this kind without more knowledge of the widths of the final state.

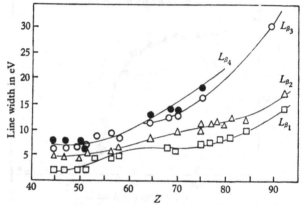

Fig. 21. Variation with atomic number of the widths of the lines $L\beta_1$, $L\beta_2$, $L\beta_3$ and $L\beta_4$ (Bril, 1947).

A puzzling feature of fig. 21 is the apparent absence of any rapid variation of widths of the $L\beta_3$ and $L\beta_4$ lines for Z between 44 and 50. In view of the interpretation of figs. 15 and 19 above, the widths of these lines should pass through maxima for Ag (47), and should decrease considerably between Ag (47) and Sb (51). A general increase of level widths with atomic number is to be expected according to a relation of the form

$$\Gamma = A + RZ^4,$$

the two terms representing the contribution from radiationless and radiative transitions respectively, but this increase should not be sufficiently rapid to mask the expected variation of L_1 width near $Z = 47$.

* Williams (1931), Richtmyer, Barnes and Ramberg (1934), Schrader (1936), Parratt (1938), Cooper (1942), Coster and Bril (1942), Bril (1947).

For $Z > 73$, on the other hand, fig. 21 shows the expected increase of line width of $L\beta_3$ and $L\beta_4$, relative to $L\beta_1$ and $L\beta_2$.

The most likely explanation of the anomalies in the L series line widths is that they are associated with irregular variations of the width of the final state due to Coster-Kronig transitions between levels in the M shell. For example, Table XIII shows that the Coster-Kronig transition $M_{III} \to M_{IV,V} N_{II}$ is probably important for Z between 50 and 65, and this may be responsible for some of the increases of the width of L_{β_3} in this region.

Irregular variations of line widths with Z have been observed for M_ζ ($M_V - N_{II,III}$) by Kiessig (1938). He found the width of this line to increase from 1·99 eV. for Sr (38), to 18·5 eV. for Te (52), and then to drop to 7·0 eV. for Ba (56). Cooper (1944) attributed this variation to the widening of the $N_{II,III}$ levels by the Coster-Kronig transition $N_{II,III} \to N_{IV,V} N_{IV,V}$, which is energetically possible for $Z < 56$–60.

The most detailed analysis of the effect of Coster-Kronig transitions on line widths has been made for gold. Pincherle (1934, 1935a, b, c) has estimated the contribution to the widths arising from radiative and Auger transitions with the results shown in Table XV.

TABLE XV. *Widths of L levels of gold*

Level	Calculated width (eV.)			Observed width (Richtmyer, Barnes and Ramberg)
	Radiation	Auger	Total	
L_I	1·0	5·5	6·5	8·7
L_{II}	0·9	2·2	3·1	3·7
L_{III}	1·6	2·6	4·2	4·4

The great importance of Coster-Kronig transitions in increasing the width of the L_1 level is clear from the table, and the agreement with the widths estimated by F. K. Richtmyer, Barnes and Ramberg (1934) (given in the last column of the table) is reasonably good.

Ramberg and F. K. Richtmyer (1937) have calculated level widths arising from a number of Coster-Kronig transitions for gold. In Table XVI they are compared with the corresponding widths observed by F. K. Richtmyer and colleagues.

TABLE XVI. *Effect of Coster-Kronig transitions in influencing line widths for gold*

Level	Calculated width (eV.)			Observed width (eV.)
	Radiation	Auger (Coster-Kronig)	Total	
L_I	1·78	11·91	> 13·69	8·7
M_I	0·078	10·23	> 10·31	15·5
M_{II}	0·09	11·49	> 11·58	10·7
M_{III}	0·05	4·45	> 4·50	12·1
N_I	0·003	13·60	> 13·60	11·7

The agreement with experiment is fair. It must be remembered, as pointed out above, that the experimental level widths are deduced from line widths which represent the sums of the widths of the initial and final states, and that the actual assignment of contributions to the line width from the two levels concerned is not entirely definite.

The large difference between the calculated values of Pincherle and of Ramberg and Richtmyer for the L_1 width is rather surprising. It may arise from Pincherle's use of hydrogenic wave functions, while Ramberg and Richtmyer used wave functions appropriate to the Fermi-Thomas field for Tl++. If the Coster-Kronig width for the L_1 shell of gold is as large as that calculated by Ramberg and Richtmyer, the intensities of the Au $L\beta_3$ and $L\beta_4$ lines relative to Au $L\beta_1$ would be expected to be considerably smaller than observed (see fig. 19). Pincherle's calculations are consistent with the observed values.*

To sum up the evidence for the importance of Coster-Kronig transitions from line widths, there can be no doubt that these transitions supply the explanation of the abnormal breadths of lines arising from an initial state of small azimuthal quantum number. Details of the variation of line widths with atomic number, however, are not understood so well as the variation with Z of diagram and satellite line intensities.

* See §4.10.

4.10. The influence of other Auger transitions on X-ray spectra

(i) X-ray satellites

So far we have concentrated attention on Coster-Kronig transitions because of their great significance in X-ray spectra. Other types of Auger transitions, however, may also be important in understanding certain aspects of X-ray spectra.

For example, every Auger process leaves an atom multiply-ionized in inner shells, and subsequent transitions will give rise to

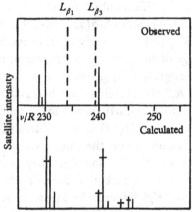

Fig. 22. Position and relative intensity of L-series satellites observed by Burbank (1939) in Ag and attributed to transitions $LL{\to}LM$, following the Auger transition $K{\to}LL$. The calculated satellite structure is due to R. D. Richtmyer (1939).

satellite lines. Burbank (1939) observed four L-series satellite lines in Ag (47) which he attributed to transitions LL-LM, the initial state arising as a result of the Auger transition K-LL. R. D. Richtmyer (1939) estimated the positions and intensities of the satellite lines to be expected from transitions in such a doubly-ionized atom. His calculated satellite pattern is compared with that observed by Burbank in fig. 22, and the detailed agreement is seen to be surprisingly good. Vieth and Kirkpatrick (1939), however, made a very careful search for similar satellites in the Mo L series but failed to detect them. It is difficult to understand why they should occur in one case and not in the other.

Satellites formed in this way would in any case be weak compared with other L-series lines, if excited by electron impact, because they depend on the ratio n_K/n_L. Under the most favourable conditions this ratio is not likely to exceed 0·05 for silver.* Such satellites, however, should be readily detected in fluorescent spectra. They do not seem to have been looked for.† The X-ray satellites so far discussed are observed on the high-frequency side of the parent line, but K and L satellites on the low-frequency side have been observed recently by Cauchois and her co-workers using the powerful technique of the curved crystal X-ray spectrometer. Thus Hulubei (1947) observed a satellite (designated $K\alpha_s$) on the low-frequency side of the $K\alpha$ doublet for elements in the range of atomic number from As (33) to Mo (42). In each case the frequency separation from $K\alpha_1$ was very nearly equal to the frequency of the M absorption edge of the element one higher in atomic number. A similar line (designated $K\beta_s$) has been observed on the low-frequency side of $K\beta_1$, by Hulubei, Cauchois and Manescu (1948).

These observations are clearly related to the much earlier observation by Bloch and Ross (1935) of a considerable increase in the intensity of the continuum on the long wave-length side of the weak $K\beta_5$ lines of Mo, Rh, Pd and Ag. They found a rise in the intensity of the continuous background over a wave-length region of about 2 X.U. from $K\beta_5$ followed by a plateau extending right up to $K\beta_1$. The radiation intensity on the plateau was practically equal to that of $K\beta_5$ itself.

Both these observations have been interpreted as a partial Auger effect or a kind of internal Raman effect, in which the parent line undergoes partial internal conversion in the $\dot{M}$ shell of the same atom. An electron in the M shell absorbs part of the radiation energy and the frequency of the radiation is decreased. Bloch (1935) made a rough calculation of the probability of the effect and found reasonable agreement with the intensity of the continuum he actually observed.

It is not clear why a satellite line rather than a continuum was observed in the experiments of Cauchois and her colleagues since

---

* Burhop (1934).

† Satellites of this kind should also be observed with appreciable intensity in the radiation emitted following K capture.

the outer electron is presumably ejected into the continuum, but it may be that the probability of the process is greatest for small energies of ejection. Cauchois (1943) also observed satellites on the long wave-length side of $L\alpha_1$ for heavy elements, but these could possibly be interpreted as Coster-Kronig satellites of $L\alpha_2$. These lines are labelled α_s in the spectra of Plate III.

(ii) *Line intensities*

The true relative intensities of X-ray lines can be calculated from (4·6). Owing to the factor n_a/n_r, the relative intensities of lines originating in different initial states will depend on the method of excitation. Very few measurements of relative intensity are available for lines excited by fluorescence. For excitation by electron impact calculated values of n_a/n_r are available, but there is some uncertainty as to the correct value to take for this quantity in calculating relative intensities because a considerable proportion of the radiation emitted by electron impact on a target actually arises from secondary excitation by the continuous radiation.* The relative intensities of the $L\alpha_1$, $L\beta_1$, $L\beta_3$ and $L\beta_4$ lines of gold have been calculated, assuming that the values for n_a/n_r corresponding to direct excitation by electron impact can be taken.† Table XVII shows the results of the calculation. Each of the factors contributing to the intensity of a line is listed in Table XVII, relative to its value for $L\beta_1$. Values of Γ_{ab}^R calculated, using relativistic quantum theory, were used,‡ while values of $\Gamma_a^R + \Gamma_a^A$ were obtained by averaging the results of Richtmyer, Barnes and Ramberg (1934) for gold and of Bearden and Snyder (1941) for platinum, giving widths of 8·3, 3·6 and 4·0 eV. for the L_I, L_{II} and L_{III} levels. The intensities of $L\beta_3$ and $L\beta_4$, relative to $L\beta_1$, are in good agreement with those observed, but the calculated intensities of $L\alpha_1$ and $L\beta_2$ are much too small.§ Too much importance, however, should not be attached to this comparison, because there are several uncertain quantities in the calculation and the limits of error in the level width determinations are rather large.

* Stoddard (1934, 1935). † Burhop (1940). ‡ Massey and Burhop (1936 b).
§ Pincherle (1935 c) obtained much better agreement when he first made this comparison. However, he seems to have used an expression for the relative intensity that would be obtained by multiplying (6) by Γ_a^R/Γ_r^R. This does not seem justified.

TABLE XVII. *Estimated relative intensities of*
Au $L\alpha_1$, $L\beta_1$, $L\beta_2$, $L\beta_3$, $L\beta_4$ *lines*
(Taking $L\beta_1$ as 1.)

	n_a	$1/(\Gamma_a^R + \Gamma_a^A)$	Γ_{ab}^R	ν_{ab}	Calculated intensity	Measured intensity
$L\alpha_1$	2·23	0·90	0·69	0·85	1·19	2·0
$L\beta_1$	1·0	1·0	1·0	1·0	1·0	1·0
$L\beta_2$	2·23	0·90	0·146	1·03	0·31	0·45
$L\beta_3$	0·96	0·43	0·32	1·02	0·135	0·13
$L\beta_4$	0·96	0·43	0·23	0·99	0·094	0·10

4.11. The role of the Auger effect in the interpretation of soft X-ray spectra of solids

(i) *The shape of the X-ray emission bands*

While for a free atom the levels of a valence electron have discrete energy values, for a solid they form a continuum. These levels are normally filled up to a certain level of energy E_m say, depending on the temperature of the solid. It is convenient to measure the energies of the levels of such a continuum by taking the energy of the lowest level as zero. Their distribution is specified by a density function $N(E)$ such that $N(E)\,dE$ is the number of levels in the range between E and $E+dE$.

In actual solids the energy E_m of the highest filled level in a valence band may range from 3 to 40 eV. After inner-shell ionization in an atom of a solid the vacancy may be filled by the transition of an electron from the valence band. The X-radiation thus emitted will not be in a discrete line, however, but will extend over a range of energies of extent E_m. For short wave-length X-radiation this energy spread is not very noticeable, particularly as it may often be small compared with the natural width of the inner level. For soft X-rays, however, such radiation will extend over an emission band and will be very different from the sharp lines characteristic of free atoms. Fig. 23 shows the shape of a typical emission band (Skinner, 1940) due to the transition of an electron from the valence (or conduction) band to the L_{III} level of sodium. The range of wave-lengths covered by this band is from 405 to 445 Å.

If ν_0 is the frequency of the radiation emitted when a valence electron at the bottom of the emission band falls to the vacant inner level, the intensity $I(\nu)\,d\nu$ of the X-radiation emitted in the frequency range between ν and $\nu + d\nu$ is given by

$$I(\nu) = k\nu^4 f(E) N(E), \qquad (4.18)$$

where $E = h(\nu - \nu_0)$, k is a constant, and $\nu^3 f(E)$ is the probability per unit time that an electron of energy E will make a transition to the vacant inner level.

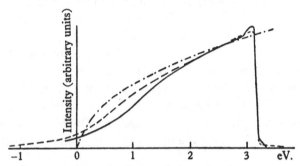

Quantum energy of radiation relative to band zero

Fig. 23. Shape of the L_{III} emission band of sodium. —— intensity distribution observed by Skinner; —·—·—· shape corresponding to an intensity distribution proportional to $E^{\frac{1}{2}}$; ---- intensity distribution calculated by Landsberg using Skinner's theory and assuming interaction between two electrons in the conduction band of the form $V(r) = \exp(-\eta r)/r$ ($\eta = 1\cdot21$ (A.)$^{-1}$).

The valence band can be regarded as a superposition of a number of bands, the electronic wave functions for which have the angular properties of $s, p, d, \ldots$ orbitals. If $N_s(E)$, $N_p(E)$, $N_d(E)$, $\ldots$ are the level densities in these bands

$$N(E) = N_s(E) + N_p(E) + N_d(E). \qquad (4.19)$$

In electron transitions to the inner shell the ordinary selection rules for optical transitions apply, and, in particular, the azimuthal quantum number must change by 1 unit. Thus only valence electrons in p levels may make transitions to the K or L_I shells, while only valence electrons in s or d levels may make transitions to the L_{II} or L_{III} shells.

Writing f_{sp}, f_{dp} for the transition probabilities for s or d valence electrons to the L_{III} shell,

$$I(E) = k\nu^4 \{f_{sp}(E) N_s(E) + f_{dp}(E) N_p(E)\}. \qquad (4.20)$$

The quantities f_{sp}, f_{dp} are nearly independent of E, while near $E = o$, $N_s(E) \propto E^{\frac{1}{2}}$, $N_p(E) \propto E^{\frac{3}{2}}$. Over the occupied range of energy levels the variation of ν^4 with E can be ignored. The emission band would therefore be expected to have a characteristic shape somewhat similar to the dot-dash curve of fig. 23. At the low-energy (long wave-length) end of the band the intensity distribution should rise sharply in proportion to $E^{\frac{1}{2}}$. At the short wave-length end of the band the intensity distribution should fall suddenly to zero* at a frequency corresponding to transitions from valence levels of energy E_m, the highest normally occupied level of the valence band.

The observed shape of the sodium L_{III} emission band (fig. 23) shows many of the expected features. At the low-energy end of the band however, the intensity, instead of going to zero like $E^{\frac{1}{2}}$, has a pronounced tail extending for a little more than 1 eV. past the expected low-energy limit of the band. Similar long wave-length tails are a prominent feature of L_{III} emission bands of other light metals.

Skinner (1940) gave the explanation of these tails. When a transition occurs between an electron in the conduction band and an unoccupied inner level, a vacancy is produced in one of the normally occupied levels of the conduction band. This state has a very short lifetime, however, since the interaction between the electrons of the conduction band will lead to an Auger transition in which a conduction electron from a higher level falls into the vacant level and the excess energy is used to raise a second conduction electron to a level of energy greater than E_m.

Owing to the very short lifetime Δt of a state in which there is a vacancy in a normally occupied level of the conduction band, this level cannot have a sharply defined energy but must have a finite breadth ΔE given by the relation $\Delta E \doteqdot \hbar/\Delta t$. Each energy level in the conduction band must then be replaced by a broadened level. Since the lifetime of a vacancy in a level at the bottom of the band will be shorter than for a vacancy near the top of the occupied levels, the spread of energy for the former state will be broader. As a result of this broadening the density function $N(E)$ in equation (4.16) has to be modified, since every elementary portion $N(E) dE$

---

* The fall will not be quite sudden owing to the temperature distribution of the upper occupied levels.

of the old distribution has now to be replaced by a portion of the same area but broadened out over a finite range of E. In particular, the distribution will not fall to zero at $E = o$, but owing to the finite breadth of the lower lying levels the conduction band will have a tail extending toward negative energies. The emission bands will show a corresponding tail. Numerical calculations based on Skinner's theory have been made by Landsberg (1949) for the Na L_{III} emission bands.

Assuming a Coulomb interaction between electrons in the conduction band he calculated the length of the long wave-length tail to be much greater than that measured by Skinner. However, for a 'screened' potential of the form $r^{-1}\exp(-\eta r)$ between two electrons at a distance r Landsberg was able to get a reasonable length for the tail (about 1 eV.) by taking a value for η of $1 \cdot 21 \times 10^8$ cm.$^{-1}$, a result which he interprets as suggesting the existence of correlations between electrons in a metal, other than those arising from the Pauli principle. The shape of the L_{III} emission band for sodium assuming the above value of η is shown in fig. 23.

Thus although Skinner's theory accounts qualitatively for the shape of the emission bands it cannot yet be said to provide a satisfactory quantitative explanation.*

(ii) *Coster-Kronig transitions in the X-ray spectra of solids*

Coster-Kronig transitions play an important part in the interpretation of intensity and breadth anomalies in the soft X-ray emission spectra of solids. For light elements the difference in ionization energy between the L_I and L_{III} shells is much smaller than for heavier elements, but, on the other hand, in the case of metals a continuum of unoccupied states is always available just above the occupied states of the conduction band. There is generally therefore no difficulty in finding a final state for which a Coster-Kronig transition $L_I \rightarrow L_{II, III} V$ ($V =$ valence electron) is possible

---

* Cady and Tomboulian (1941) have sought to explain the long wave-length tail of soft X-ray emission bands as a partial Auger effect similar to that suggested by Bloch to explain the continuum on the long wave-length side of the $K\beta_5$ line of Mo (see §4.10). They suppose the long wave-length radiation to arise in a double transition in which one conduction electron falls to the vacant inner level, while another is raised to a level above the top of the levels normally occupied in the conduction band. No detailed calculations have been made on this theory, which seems less attractive than that proposed by Skinner.

with a high probability. Even for insulators such as phosphorus or sulphur, where there are no conduction levels available, the energy gap between the filled valence bands and the lowest excited bands is of the order of a few electron-volts. Since the difference in energy between atoms ionized in the L_I and L_{III} shells is about 10 eV., Coster-Kronig transitions are still energetically possible.

Owing to the existence of the vacant conduction levels, transitions of the type $L_{II} \rightarrow L_{III} V$ are also energetically possible for a metal, even for sodium, where the energy difference between the L_{II} and L_{III} levels is only 0·2 eV. This type of transition is not, however, usually possible for insulators. For metals, the Coster-Kronig transition $L_I \rightarrow L_{III} V$ seems to have a much higher transition rate than is the case for the corresponding transition in a free atom. Thus Skinner (1940) was unable to detect the L_I emission bands for Na, Mg and Al metals, although he should have done so readily if their intensity had been even 1 % of the intensity of the L_{III} emission band. At the same time the L_I absorption edges of Mg and Al metals are very diffuse.

Although it is not possible to detect the L_I emission band of Al directly, Skinner pointed out that its width can be inferred from some measurements of K satellite emission bands. Karlsson and Siegbahn (1934) detected such bands in the K spectrum of Al and attributed them to transitions $KL_I \rightarrow VL_I$ and $KL_{II\,III} \rightarrow VL_{II,III}$, i.e. to transitions of a valence electron to the K shell of an atom also ionized in an L shell. Since the lifetime of a state of K ionization is much less than for L ionization, radiation can occur from an atom doubly ionized in the K and L_I shells before there is time for the L_I shell to be filled by a Coster-Kronig transition.

From the shape of the short wave-length side of these emission bands it is possible to estimate the width of the corresponding state of inner ionization just as was done for absorption bands (§ 4.9). The widths of the KL_I and $KL_{II,III}$ states of double ionization estimated in this way were found to be 2·5 and 0·5 eV. respectively. Evidently then the Coster-Kronig transitions broaden the L_I level by about 2·0 eV. relative to L_{III}. Since the width of the Al L_{III} state as determined from absorption-edge measurements is much less than 0·1 eV., the failure to observe the L_I emission band in the cases mentioned above is readily understood.

The probability of the transition $L_{II} \rightarrow L_{III} V$ is apparently smaller than that of the $L_I \rightarrow L_{III} V$ transition. Thus the L_{II} emission bands of Na, Mg and Al, although weak, have been observed. In the absence of Coster-Kronig transitions the relative intensity of the L_{III} to the L_{II} emission bands should be 2. Skinner observed values of this ratio of 50 for Na, 20 for Mg, 9 for Al, 3 for Si, 2 for P and S and 10 for Ca and Cu.

The normal value obtained for the insulators P and S as predicted above, in contrast to the high values for metals, provides convincing evidence of the importance of Coster-Kronig transitions in the interpretation of these soft X-ray emission spectra.

THE INTERNAL CONVERSION OF γ-RAYS

5.1. The internal conversion coefficient

The internal conversion coefficient (I.C.C.) of a γ-radiation has been defined in Chapter II as the ratio, α, of the number of conversion electrons, n_e, to the number of γ-ray quanta, n_ν, emitted per unit time, i.e. $\alpha = n_e/n_\nu$.

The I.C.C. α may be regarded as made up of a number of 'partial' I.C.C.'s, α_K, α_L, α_M, corresponding to conversion in the various electron shells. Thus $\alpha_K = n_e^K/n_\nu$, where n_e^K is the number of electrons ejected per unit time from the K shell.

Clearly
$$\alpha = \alpha_K + \alpha_L + \alpha_M + \dots \qquad (5.1)$$

If a nucleus emits γ-rays of more than one quantum energy there will in general be a different value of α for each quantum energy. As pointed out in Chapter II, the calculation of the transition rate for internal conversion of γ-radiation follows lines very similar to those traced out for the Auger effect. In the γ-ray case, however, the question of indistinguishability of particles does not arise. On the other hand, there is no unique nuclear model from which one can write down proper wave functions of the nuclear particles taking part in the transition, and, *a priori*, the type of transition involved is not known. As a result, in the calculation of the transition rate b_n using expression (2.9) it is uncertain what form should be used for the potentials A_0, $\mathbf{A}$ occurring in the interaction (2.8).

5.2. Radiation from electric and magnetic multipoles

For an arbitrary distribution of charges and currents the potentials A_0, $\mathbf{A}$, which define the radiation field, are given by equations (2.5), (2.6).

Since the wave-length of any of the radiations considered is much greater than nuclear dimensions,

$$2\pi\nu_{fi}r'/c\,(=qr') \ll 1,$$

and
$$\exp(-iqr'\cos\theta) \doteq 1 - iqr'\cos\theta. \qquad (5.2)$$

Further, taking the centre of the nucleus as origin, to calculate the potentials at some distance r from the nucleus ($r \gg r'$), write

$$|\mathbf{r}-\mathbf{r}'| \doteqdot r - r' \cos\theta, \qquad (5.3)$$

$$1/|\mathbf{r}-\mathbf{r}'| \doteqdot 1/r + (r' \cos\theta)/r^2. \qquad (5.4)$$

Using (5.2), (5.3), (5.4) in equation (2.6) of Chapter II,

$$\left.\begin{aligned} a_0 &\doteqdot \{\exp(iqr)/r\} \int \rho_{fi}\, d\mathbf{r}' - \{i\exp(iqr)/r^2\}\,(i+qr) \int \rho_{fi}\, r'\cos\theta\, d\mathbf{r}', \\ \mathbf{a} &\doteqdot \{\exp(iqr)/cr\} \int \mathbf{j}_{fi}\, d\mathbf{r}' - \{i\exp(iqr)/cr^2\}\,(i+qr) \int \mathbf{j}_{fi}\, r'\cos\theta\, d\mathbf{r}'. \end{aligned}\right\}$$

$$(5.5)$$

The first terms in these expansions for a_0, $\mathbf{a}$ represent the potentials that would be produced by an electric dipole placed at the centre of the nucleus.† The second terms represent a field that would be produced by a combination of an electric quadrupole and a magnetic dipole placed at the origin.‡

Similarly, if the approximation of equations (5.2), (5.3), (5.4) had been continued to the next order, third terms would have been obtained in the expansions of (5.5) representing a combination of the radiation emitted by an electric octopole and a magnetic quadrupole. If the expansion were to be continued indefinitely the lth terms would represent a mixture of electric 2^l-pole and magnetic 2^{l-1}-pole radiation.

Successive terms in the expansion (5.5) in general diminish approximately in the ratio qr_0, where r_0 is the nuclear radius. It may happen, however, that the first few terms of the expansions vanish owing to the vanishing of the charge and current integrals corresponding to certain nuclear transitions. It is clear that the more terms that vanish in this way the weaker the radiation field.

The expressions to be used for ρ_{fi}, $\mathbf{j}_{fi}$ in (5.5) depend on the particular nuclear model assumed. It is therefore not possible, without assuming some particular model, to calculate the absolute

---

† Actually the first term in the expansion for a_0 vanishes since $\int \rho_{fi}\, d\mathbf{r}' = 0$ for the radiation resulting from a transition between two stationary states.

‡ For the proof that the field given by the second terms can be considered as compounded of the fields due to a magnetic dipole and an electric quadrupole, see, for example, Stratton (1941, p. 433).

transition rate in the nucleus, and thence the lifetime of an excited nuclear state. However, calculations of the internal conversion coefficient for γ-rays have been carried out assuming a small radiator located at the centre of the nucleus and emitting a field characteristic of some particular type of radiation (electric or magnetic dipole or quadrupole, etc.). The appropriate scalar and vector potentials corresponding to the assumed type of radiation are substituted in equation (2.9) to give the transition probability for internal conversion of the radiation with the electronic transition $\chi_i \to \chi_f$. The potentials corresponding to some of the simpler types of radiation field are given in Chapter II, equations (2.19), (2.20), (2.21).†

5.3. Selection rules in nuclear transitions

The charge and current densities ρ_{fi}, $\mathbf{j}_{fi}$ can be written in terms of the initial and final nuclear wave functions ψ_i, ψ_f. Although the detailed forms of ψ_i, ψ_f depend on the assumed nuclear model, without any such model it is possible to derive certain selection rules. Let J, J' be the total nuclear angular momentum quantum numbers in the initial and final states. It was shown by Heitler (1936)‡ that a quantum of 2^l-pole electric or magnetic radiation can be regarded as possessing an angular momentum $l\hbar$. The conservation of angular momentum requires therefore that

$$|J-J'| \leqslant l \leqslant |J+J'|. \tag{5.6}$$

This means that in the expansion (5.5) the only non-vanishing terms are those corresponding to 2^l-pole electric or magnetic radiation with l satisfying the inequality (5.6).

Owing to the rapid diminishing of successive terms in the expansion (5.5) the most important contribution to the radiation field will generally come from the terms that correspond to the smallest permitted value of l, viz. $l = |J-J'|$. Since magnetic 2^l-pole radiation occurs in (5.5) in association with electric 2^{l+1}-pole radiation, it would be expected that 2^l-pole electric radiation ($l = |J-J'|$)

† The forms given there are not the only ones. For a radiation of polarity 2^l there are $2l+1$ independent forms of the potentials. The calculated transition rate depends on the form assumed, but the internal conversion coefficient does not.

‡ See also Berestetzky (1947).

would provide the most important contribution to the radiation field.

This conclusion is modified, however, by a further selection rule provided by the parity properties of the nuclear wave functions. If the signs of all the spatial coordinates of the nuclear particles are changed, the corresponding state is odd or even according as the nuclear wave function changes sign or not. The matrix elements which determine the rate of emission of multipole radiation are of the form

$$\int \psi_f^* M \psi_i \, d\tau, \qquad (5.7)$$

where M is an operator depending on the multipole nature. Integrals such as (5.7) vanish if the integrand is an odd function of $\cos\theta$. The operator M is odd or even for 2^l-pole electric radiation according as l is odd or even. For 2^l-pole magnetic radiation it is odd when l is even and even when l is odd. It follows that electric 2^l-pole radiation can occur only as a result of transitions between states of the same parity if l is even, and only between states of different parity if l is odd. Just the opposite applies for magnetic 2^l-pole radiation. If electric 2^l-pole radiation can occur $(l=|J-J'|)$ the transition is said to be parity allowed; if not, parity forbidden. If then electric 2^l-pole radiation $(l=|J-J'|)$ is forbidden by the parity selection rule, the radiation field will consist mainly of a mixture of magnetic 2^l-pole and electric 2^{l+1}-pole radiation.

Table XVIII shows how the selection rules apply in different cases.

Of course larger values of l up to $|J+J'|$ may also contribute to the radiation field, but because of the smallness of the ratio qr_0 between successive terms in (5.5) the lowest permitted value of l will generally determine the nature of the radiation field.

TABLE XVIII. *Minimum allowed multipole orders*

$=|J-J'|$	Parity change?	Electric	Magnetic
Odd	Yes	2^l-pole	2^{l+1}-pole
Odd	No	2^{l+1}-pole	2^l-pole
Even	Yes	2^{l+1}-pole	2^l-pole
Even	No	2^l-pole	2^{l+1}-pole

In addition to the above rules there is the further rule that the radiative transition $J'=0 \to J=0$ is entirely forbidden.

5.4. Calculation of the type of radiation emitted, using different nuclear models

In the expansion of (5.5) it is seen that 2^l-pole magnetic radiation is associated in the same term with 2^{l+1}-pole electric radiation. If, therefore, the minimum allowed order of magnetic multipole radiation is 2^l-pole, its intensity should be comparable with that of the corresponding electric multipole radiation (2^{l+1}-pole). On the other hand, if the minimum allowed magnetic multipole radiation is 2^{l+1}-pole, its intensity should be very much smaller than that of the 2^l-pole electric radiation which is permitted in this case.

The actual relative amplitudes of magnetic 2^l-pole and electric 2^{l+1}-pole radiations depend on the assumed nuclear model. The nature and rate of radiation have been investigated using a number of different nuclear models. The simplest was that of Weiszäcker (1936), who considered the nuclear electromagnetic radiation to be produced by a dipole oscillator of charge Ze, whose amplitude of oscillation was equal to the nuclear radius r_0. The transition rate, p, for electric 2^l-pole radiation was calculated to be

$$p \doteq (Z^2/137)(E/\hbar)(Er_0/\hbar c)^{2l}, \qquad (5.8)$$

where E is the energy of the emitted quantum.

Flügge (1941) made a similar calculation considering the nucleus to be a rotator of mass M, moment of inertia I, and obtained for the transition rate of 2^l-pole electric radiation,

$$p \doteq (Z^2/137)(\hbar/I)(E/Mc^2)(Er_0/\hbar c)^{2l}, \qquad (5.9)$$

which gives, assuming reasonable values of I, a result much smaller than that given by (5.8).

Fisk and Taylor (1934) studied a crude model of a heavy nucleus in which a proton was considered to move in a potential well with steep walls ($> 2Mc^2$ deep) and width 10^{-12} cm. For transitions corresponding to a change in angular momentum of two units they found a ratio of the amplitudes of the electric quadrupole and magnetic dipole fields of about 4000:1.

On the other hand, Hulme, Mott, F. Oppenheimer and Taylor (1936), assuming a nuclear model in which a proton moved in a central Coulomb field of strength adjusted to give proton orbits of radii 10^{-12} cm., found that for transitions of the type $\Delta J = 1$ but involving no change in the orbital angular momentum (i.e. no parity change) the magnetic dipole field amplitudes were larger than the electric quadrupole amplitudes at the longer wave-lengths, the ratio of the two being roughly proportional to the wave-length.

A somewhat similar model was employed by Koyenuma (1941), while Hebb and Uhlenbeck (1938) assumed the radiating particle to be a single α-particle.

Several authors have investigated the radiation from a liquid-drop nuclear model. Bethe (1937) calculated the rate of radiation on the basis of a liquid-drop model, using a classical theory to calculate the rate at which energy was radiated as a result of the currents caused by surface vibrations. He obtained for the transition rate for electric 2^l-pole radiation

$$p = (K/l!)^2 (e^2/\hbar r_0) (Er_0/\hbar c)^{2l+1}, \qquad (5.10)$$

where K is a constant of the order of magnitude of unity. Flügge (1941), Fierz (1943) and Berthelot (1944) have also investigated this model, and the latter gives for the transition rate for 2^l-pole electric radiation

$$p = \frac{3(l+1)}{1^2 \cdot 3^2 \cdot \ldots \cdot (2l+1)^2} \left(\frac{Z^2 e^2}{Mcr_0^2}\right) \left(\frac{Er_0}{\hbar c}\right)^{2l}. \qquad (5.11)$$

For 2^l-pole magnetic radiation the corresponding expression is obtained by replacing l by $l+1$ in (5.11).

The uncertainty as to what constitutes a suitable nuclear model scarcely justifies further speculation along these lines. None of the models already used gives a good account of the observed transition rates from nuclear excited states, but it is clear that practically any ratio of magnetic to electric radiation could be accounted for by an appropriate choice of model.

There is one other aspect that should be mentioned at this stage. It is well known (see Bethe, 1937) that transition rates for nuclear electric dipole radiations are much smaller than expected, and in fact are generally of the same order as those for electric quadrupole radiation, contrary to the conclusion which might be drawn from the expansion of (5.5) above. If the centre of mass and centre of charge of the nucleus coincide, the dipole moment would vanish whatever the other details of the nuclear structure. In actual nuclei these two centres almost certainly do not exactly coincide, but their distance part is probably small; thus the nuclear electrical dipole transition rates are also small. No such effect is to be expected for other types of multipole radiation, whether electric or magnetic.

5.5. The measurement of the internal conversion coefficient

The quantities of physical interest that may be determined experimentally for any γ-ray are the total I.C.C. α, and the partial I.C.C.'s α_K, α_L, ... If the absolute values of these partial I.C.C.'s cannot be measured, their ratios may still be of interest. From these quantities information can be obtained about the angular momentum change associated with the corresponding nuclear transition.

The determination of α requires a measurement of the emission rates both of the γ-ray quanta responsible for the conversion and also of the conversion electrons.

We consider first the simplest possible case in which a γ-radiation of a single quantum energy is emitted from a nucleus and is not accompanied by any other primary α- or β-radiation. Such a state of affairs may be encountered in the γ-ray emission from the isomer of a stable nucleus (e.g. ^{87}Sr, ^{115}In). In this case the most direct method of measuring α consists in comparing the amounts of ionization produced in a chamber by conversion electrons and by their associated γ-rays. The ionization due to the electrons can be separated from that due to the γ-radiation by placing a suitable absorber between the source and the ionization chamber. If the length of the chamber is d (less than the range, r, of the electrons in the gas of the chamber), E_e and E_γ the energy of the electrons and the quantum energy of the γ-radiation respectively, and μ_γ the absorption coefficient of the γ-rays in the gas of the chamber, the ratio of the ionization due to electrons to that due to γ-radiation is (approximately)

$$R_{e,\gamma} = \alpha E_e d / r E_\gamma \{ 1 - \exp(-\mu_\gamma d) \}. \tag{5.12}$$

This relation assumes that the quantum energy of the radiation greatly exceeds the inner-shell ionization energy of the shells in which the internal conversion occurs, so that a mean energy E_e can be taken for all conversion electrons.

Alternatively, instead of measuring the ionization produced by the conversion electrons, that due to the X-radiation emitted in the reorganization of the electron shells following internal conversion may be measured.

The ratio of the ionization produced by this radiation to that produced by the γ-radiation is (approximately)

$$R_{X,\gamma} = (\varpi\alpha E_X/E_\gamma)\{\mathrm{I} - \exp(-\mu_X d)\}/\{\mathrm{I} - \exp(-\mu_\gamma d)\},$$

$$(5.13)$$

where E_X is the (mean) quantum energy of the emitted X-radiation, μ_X is its absorption coefficient in the gas of the ionization chamber and ϖ its fluorescence yield. Since K, L, K, ... series radiation can be readily separated by differential absorption, this method enables an estimate to be made of the partial i.c.c.'s in different shells.

If the internal conversion follows K capture, α can be calculated by measuring the intensity of the K-series radiation emitted from the source relative to that of the γ-rays or conversion electrons. If ϖ_K is the fluorescence yield of K-series radiation there will be $\alpha/(\mathrm{I}+\alpha)$ conversion electrons and $\varpi\{\mathrm{I}+\alpha_K/(\mathrm{I}+\alpha)\}$ K-series quanta per disintegration. The ionization produced by electrons, by K series quanta and by γ-rays, will then be in the ratio

$$E_e\alpha d/r:\varpi_K(\mathrm{I}+\alpha+\alpha_K)E_X\{\mathrm{I}-\exp(-\mu_X d)\}:E_\gamma\{\mathrm{I}-\exp(-\mu_\gamma d)\}.$$

$$(5.14)$$

If Geiger counters instead of ionization chambers are used to determine α, expressions (5.12)–(5.14) can still be used, but since each electron or measured photoelectron will produce a single pulse in the counter, $E_e d/r$, E_X and E_γ must be replaced by 1.

In all these expressions appropriate corrections must of course be made for absorption in the window of the counter, for wall corrections, and for the error made in assuming a uniform rate of energy loss of the electron.

We consider next the measurement of α in the simplest possible case of a radioactive transformation. We suppose that in the disintegration there is 100 % excitation of a single level from which the only possible transition is to the ground state, so that the total number of primary α (or β) particles ejected can be taken to give the total number of transitions. If then the number of conversion electrons or of X-ray quanta following internal conversion can be measured, α, α_K, α_L, ... can be determined.

In cases of this kind, however, the i.c.c.'s can be determined more elegantly by using coincidence counting techniques. Suppose a β-ray emission is always associated with the transition, and that coincidences are measured between two β-particle counters. Let Ω_1, Ω_2 be the fractions of a complete sphere subtended by the two counter windows at the source, and η, η' the efficiency of either counter for the γ-rays that undergo conversion and for electrons respectively. Then the coincidence rate between β-particles and conversion electrons,

$$N_{\beta,e} = 2\alpha N\Omega_1\Omega_2\eta'^2/(1+\alpha), \qquad (5.15)$$

and the β-particle counting rate of counter (i) alone is

$$N_\beta = N\Omega_1\eta'(1+2\alpha)/(1+\alpha), \qquad (5.16)$$

where N is the number of disintegrations per unit time.

An aluminium plug is now placed in front of counter (i) and all the β-particles and conversion electrons are absorbed. The coincidence rate between the β-particles and the γ-rays is then determined, viz.

$$N_{\beta\gamma} = N\Omega_1\Omega_2\eta\eta'/(1+\alpha). \qquad (5.17)$$

The counting rate of counter (i) for γ-rays is

$$N_\gamma = N\Omega_1\eta/(1+\alpha). \qquad (5.18)$$

Then the ratio

$$(N_{\beta e}/N_\beta)/(N_{\beta\gamma}/N_\gamma) = 2\alpha/(1+2\alpha), \qquad (5.19)$$

so that α can be determined. If $\alpha\Omega_1$ is not small a correction has to be applied to allow for the possibility that a β-particle and a conversion electron may pass simultaneously through counter (1) and be recorded as a single particle.*

In most cases the γ-ray spectrum emitted during radioactive transformation is complex. Several γ-ray quanta may be emitted in cascade, and there may be alternative modes of disintegration or of transition to the ground state of the product nuclei. In such cases it is difficult to correlate the various groups of conversion electrons with their associated γ-radiations or to determine the relative numbers of nuclei that follow the alternative modes of transition to the ground state. The simple methods outlined above

---

* A modified coincidence counting procedure for the determination of α is described by Mitchell (1948).

are not applicable without further information obtained in supplementary experiments. The use of magnetic analysis of the conversion electrons is practically indispensable in the correlation of the electron and γ-ray spectra from a nucleus.

Even in the simple cases discussed above in which a single transition can occur to the ground state, magnetic analysis makes the determination of α much more definite by enabling the electrons produced by conversion in the different shells to be resolved. Magnetic analysis of the β-ray spectrum also enables a more reliable estimate to be made of the total number of β-particles emitted in a radioactive transformation when this provides the most convenient method of determining the total number of transitions between two nuclear states.

The technique of using the semicircular focusing magnetic spectrograph was developed to a high standard for naturally radioactive sources at an early date. Intense, but very thin, sources of this kind are often available (in some cases only a few atoms thick). There is then very little scattering in the source, and very sharp 'lines' can be obtained corresponding to groups of electrons homogeneous in energy. With artificially radioactive isotopes, however, it is not usually possible to obtain such intense thin sources, and in these circumstances, owing to loss of energy of the electrons in the material of the source, the lines due to internal conversion are not so sharp. To obtain reasonably sharp lines in such a case it is necessary to separate the active element chemically from the bulk of the target after irradiation. If the active material is chemically identical with the target material, separation is obviously more difficult and is often impossible.

A typical modern magnetic spectrograph, employing semicircular focusing and designed particularly for use with naturally radioactive sources, has been described by Martin, Richardson and Hsü (1948).

The use of coincidence counting techniques in association with a magnetic spectrometer, first suggested by Feather (1940), provides a powerful method of elucidating complicated decay schemes and of determining internal conversion coefficients when the γ-ray spectrum is not simple. An interesting semicircular focusing double-magnetic spectrograph has been developed by Feather,

Kyles and Pringle (1948) for investigations of this kind. The spectrograph, which uses a permanent magnet to enable the field to be maintained constant over the long periods necessary for coincidence counting, is illustrated in fig. 24. The source A is situated midway between two slits S_1 and S_2, which define two sheaves of electrons emitted in opposite directions. The two sheaves are focused by the transverse magnetic field on to the windows of the counters C_1 and C_2. The counters are mounted on two carriages

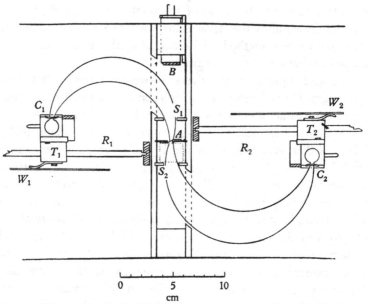

Fig. 24. Illustrating the double β-ray spectrometer suitable for coincidence measurements, developed by Feather, Kyles and Pringle (1948).

T_1 and T_2, which can be moved independently along the rods R_1 and R_2. In this way the radius of curvature of either sheaf of electrons can be varied between 2·5 and 6 cm. Electrical connexions to each counter are made by two spring strips attached to the counter carriage. One strip makes contact with the wall of the vacuum chamber and provides an earth connexion. The other makes sliding contact with an insulated copper wire (W_1, W_2) which supplies the anode voltage. Electrons can be cut off from C_1 by means of the shutter B, which can be moved down to cover the slit.

The counter can be moved from outside the vacuum chamber by rods passing through ground joints.

With this spectrometer it is possible to measure coincidences between homogeneous groups of conversion electrons and disintegration electrons of the β-ray spectrum.

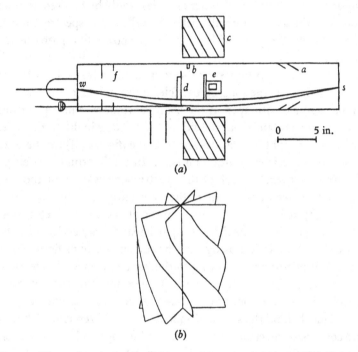

(a)

(b)

Fig. 25. Illustrating the longitudinal magnetic spectrograph used by Deutsch and his associates. (a) Diagram of the spectrograph; (b) baffle used to separate electrons and positrons.

In many cases the intensities of artificially produced radioactive sources are inconveniently low for use with the semicircular focusing magnetic spectrograph.* In such cases a considerably larger proportion of the electrons emitted in a given energy range can be focused on a counter window in a magnetic spectrograph employing longitudinal focusing. A typical lens spectrograph of this type has been developed by Deutsch, Elliott and Evans (1944).

* A novel type of magnetic spectrograph employing a transverse field and yet focusing an unusually large proportion of the electrons emitted from the source has recently been developed by Rae (1950).

It can focus electrons ejected from the source over a solid angle of 0·1 stere-radian.

Fig. 25 (a) shows diagrammatically the principle of this instrument. Electrons of some particular energy emitted by the source s are focused by the magnetic field of the coil on to the counter windows w. Electrons of other energies would be focused at other points on the axes,† but will strike the walls of the spectrograph or baffles, a, b, d, e, f, provided for this purpose. The paths of the particles in the spectrograph are helices described on conical-shaped surfaces with apex, at the source. After refraction of their paths by the lens the focused particles move in helices on a second similar surface with apex at the counter window. Since positrons and electrons rotate about the axis of the coil lens in different senses during their flight, a 'paddle-wheel' baffle (fig. 25 (b)) can be used to separate particles of different sign. In the instrument used by Deutsch the overall length of the spectrometer was about 100 cm., and the half-maximum width of the transmitted distribution varied from 0·017 to 0·06 of the momentum of the focused electrons. About 1 % of all the particles of a particular energy emitted by the source can be focused using this instrument—a high figure for a magnetic spectrograph. Owing to the high efficiency of detection, and the large distance from source to detector, this spectrograph is well adapted for counting in coincidence with a second counter placed just behind the source. Coincidence measurements of β-rays and conversion electrons with γ-rays and X-rays and of conversion electrons with β-rays have been carried out.

The easiest measurement to make with a magnetic spectrograph is that of the ratios of the partial internal conversion coefficients, α_K/α_L, α_K/α_M, α_L/α_M, ...

The measurement of the absolute values of α, α_K, α_L, ... is much more difficult because the total nuclear transition rate must be known. If the nuclear transition is always accompanied by β-decay, N can be determined by means of calibrated counters, with windows subtending a well-defined angle at the source. In this method the estimated transition rate $N_{app.}$ is less than the true rate N owing to absorption of the low-energy electrons of the β-ray spectrum in the

† Actually, with a lens spectrograph a point source is focused as a ring on the focal plane owing to the aberrations.

counter window. The correction for the window absorption can conveniently be made by determining the shape of the β-ray spectrum using a magnetic spectrograph and representing the results in the form of a Fermi plot of the function $\{A(E)/pEF(Z,E)\}^{\frac{1}{2}}$ against the energy E of the β-electrons. In this function $A(E)\,dE$ is the number of electrons detected with kinetic energy between E and $E+dE$, p the corresponding momentum and

$$F(Z,E) = 4(2pr_0)^{2s-2}$$
$$\times \exp{(\pi ZE/137p)} \,|\, \Gamma(S+iZE/137p)\,|^2/\{\Gamma(2s+1)\}^2, \quad (5.20)$$

in which r_0 is the nuclear radius, and $S = \{1-(Z/137)^2\}^{\frac{1}{2}}$.

If the β-ray spectrum is produced by a single transition the Fermi plot should be linear provided the transition is an allowed one. Usually in such a case it is found that the plot is linear over a considerable range of electron energies but departs from linearity at low energies owing to window absorption. A typical Fermi plot for the β-spectrum of ^{134}Cs is shown in fig. 26. The effect of window absorption at low energies is apparent. By extrapolating the linear part of the Fermi plot to low energies the true form of the spectrum can be obtained. Let $A_F(E)\,dE$ be the number of electrons in the energy range between E and $E+dE$ that would have been detected in the absence of window absorption. Then

$$N = N_{\text{app.}} \int A_F(E)\,dE \Big/ \int A(E)\,dE). \quad (5.21)$$

The rate of emission of conversion electrons, and thence α can be determined if the transmission factor t of the spectrometer is known. This is defined as the fraction of the emitted particles in a given energy range detected by the spectrometer counter.

If n_K is the counting rate at the peak of the line corresponding to internal conversion in the K shell,

$$\alpha_K = r_K \alpha = n_K/N_\gamma t = n_K(1+\alpha)/Nt,$$

where r_K, the proportion of internal conversion in the K shell, can also be determined from the observed internal conversion spectrum.

The I.C.C. α is then given by

$$\alpha = [(Ntr_K/n_K) - 1]^{-1}. \quad (5.22)$$

To determine the transmission t, the line width w of the spectrometer has to be measured. This is defined as the ratio of the area to the maximum height of an internal conversion line observed by means of the spectrometer.

Then assuming t to be independent of the energy of the electrons, it is given by

$$t = \int A(E)\,dE/wN_{\text{app.}}. \qquad (5.23)$$

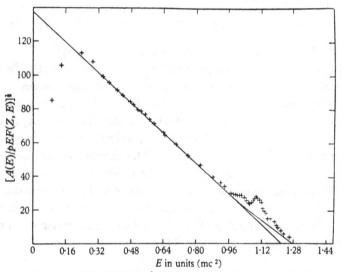

Fig. 26. Plot of $[A(E)/pEF(Z, E)]^{\frac{1}{2}}$ against E (Fermi plot) for the β-decay of ^{134}Cs. The effect of window absorption at low energies is evident (Waggoner, Moon and Roberts, 1950).

Using this method, Waggoner, Moon and Roberts (1950) estimate that in suitable cases it is possible to measure α to an accuracy of about 3 %. The measurements of very small values of α (as low as 10^{-5}) are well within the range of the available techniques.

If the nuclear disintegration scheme is complicated with several possible modes of excitation of the product nucleus and alternative modes of transition to its ground state, the I.C.C. corresponding to any particular γ-ray may be very difficult indeed to determine. The decay scheme for the transformation has to be deduced, and the fraction of all the nuclear transitions that lead to the emission of each γ-radiation determined separately. An example of the type of procedure that can lead to the determination of this fraction has already been given in Chapter III, §3.6, in connexion with the measurement by Steffen, Huber and Humbel (1949) of the fluorescence yield of platinum.

To determine internal conversion coefficients in such cases when the γ-ray spectrum emitted by the source is complex, the most suitable method consists in measuring the spectrum of the photoelectrons ejected by the γ-rays from a plate placed in front of the source. If the plate is of a material with atomic number close to that of the source, a comparison of the conversion and photoelectron spectra, together with a knowledge of the photoelectric absorption coefficients of the γ-radiation in the plate, enables α_K, α_L, α_M to be determined for the various γ-radiations. This method was used by Ellis and Aston (1930) in their study of the internal conversion spectrum of Ra C, and recently it has been applied by Martin and Richardson (1950) to the measurement of the intensities of the γ-radiations from Th$(B+C+C'')$. In order to obtain a convenient counting rate of photoelectrons a 'thick' target has to be used so that the lines observed in the spectrometer are not as sharp as those of the conversion spectrum of Th$(B+C+C'')$ from a thin source.

Fig. 27 (b) shows the photoelectron spectrum of Rd Th and its products from a gold target, while fig. 27 (a) shows the conversion spectrum of the same substance obtained by Flammersfeld (1939).

Allowing for the scattering of the photoelectrons in the 'thick' target, Richardson (1950) obtained the following relation between the height h_K of the line from the K shell in the photoelectron spectrum and the intensities I^γ in quanta per second of the γ-radiation producing it:

$$h_K = A I^\gamma \tau_K H\rho\beta^3/nZ[\log\{Rm^2c^4\beta^4/I^2Z^2(1-\beta^2)^{\frac{3}{2}}\}+1-\beta^2],$$
$$(5.24)$$

in which n is the number of atoms per cm.^{3} in the target of atomic number Z, I is the mean ionization potential of the target atoms $(\doteqdot 13\cdot 5 \text{ eV.})$, $H\rho e/c$ is the momentum of a photoelectron of velocity $v = \beta c$, $R = \Delta(H\rho)_0/H\rho$, where $\Delta(H\rho)_0$ is the distance in the spectrum between the peak of a line and its high-energy limit, and τ_K is the atomic absorption coefficient for the K shell and A is a constant.

This method assumes that the angular distribution of the ejected photoelectrons is independent of their energy. This is certainly not the case, and the variation of the angular distribution with energy will depend on the target thickness. As an example of the order of magnitude of this effect, Martin and Richardson (1950) estimated

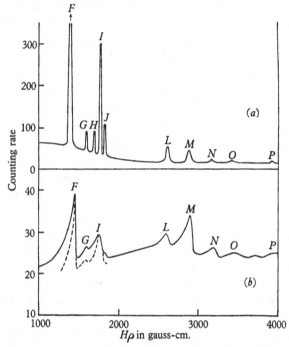

Fig. 27. Comparison between the internal conversion and photoelectron spectra of Th (B + C + C″). (a) internal conversion spectrum measured by Flammersfeld (1939); (b) photoelectron spectrum measured with gold absorber by Martin and Richardson (1950). —— gold thickness 200 mg./cm.²; ---- gold thickness 3·7 mg./cm.².

that the relative intensity of the 0·238 and 2·62 MeV. γ-radiations from the Th (B + C + C″) complex determined from the photo-electric spectrum from a gold target was about 30 % too small.

Using equation (5.24) and making a rough allowance for the variation of the angular distribution of ejection with energy, Martin and Richardson calculated the intensities of the γ-radiations from Th (B + C + C″) relative to the 2·62 MeV. radiation which is believed to occur in each disintegration to the ^{208}Pb nucleus.

Then, knowing the number of conversion electrons per disintegration for each of the lines which had been determined in previous experiments (Flammersfeld, 1939; Martin and Richardson, 1948; Feather, Kyles and Pringle, 1948), the internal conversion coefficients α_K, α_L, ... for all the lines of the spectrum could be estimated.*

The proportional counter-technique of Curran, Angus and A. L. Cockcroft (1948) appears to be capable of development into a useful method of determining i.c.c.'s in cases where details of the decay scheme have been worked out from coincidence counting measurements. Using proportional counters in a magnetic field in the manner described already in §3.2, West and Rothwell (1950) have measured α for the 37 keV. γ-radiation of ^{80}Br.

In a few cases it is possible to obtain an estimate of α by a chemical method. After internal conversion a vacancy is left in an inner electron shell. By successive transitions this vacancy is ultimately shifted to the valence electrons. If the atom is in chemical combination this loss of a valence electron may break the chemical bond, and molecules in which atoms have undergone internal conversion may be decomposed. When this happens it is possible to separate the 'activated' atoms chemically by a method very similar to the Szilard-Chalmers (1934) method of separating atoms that have undergone neutron capture. If the activated atoms (and those in which the same nuclear transition has occurred without internal conversion) are radioactive, the fraction p of the total activity that can be separated out chemically provides a measure of the i.c.c. α. In fact $\alpha = p/(1-p)$.

De Vault and Libby (1940) have found a value of p of 0·95 for the 49 keV., 4·5 hr. γ-radiation from ^{80}Br. This would lead to a value of $\alpha \doteqdot 20$, which is consistent with other data (see Table XX). This method is discussed at greater length by Segre and Helmholz (1949).

5.6. Internal conversion and the classification of nuclear energy levels

A study of the process of internal conversion has added greatly to the knowledge available for the interpretation of certain features of nuclear structure and has been of particular importance in the

---

* Itoh and Watase (1941) and Latyshev (1947) have estimated γ-ray line intensities, using Compton recoils instead of photoelectron spectra.

assignment of total angular momenta to nuclear levels. The absolute value of α_K, the K shell internal conversion coefficient depends markedly on the type (electric or magnetic) and multipole order of the γ-radiation. And, as discussed in §2.2, the multipole order of the γ-radiation depends on the change of nuclear angular momentum associated with it. For instance, a γ-radiation arising from a transition in which the nuclear angular momentum changes by units will be mainly either 2^l-pole electric or a mixture of 2^l-pole magnetic and 2^{l+1}-pole electric, according to the parity change associated with the transition.

Similarly, the ratio α_L/α_K, which is usually easier to measure than α_K, depends markedly on the angular momentum change associated with the transition. If then the angular momentum of the final state involved in the nuclear transition can be measured in some other way, as is possible, for instance, if the final state is stable, the angular momentum of the initial state can be assigned.

Internal conversion studies are especially of use in the assignment of an angular momentum to levels responsible for nuclear isomerism. Two nuclei, each of fairly long lifetime, are said to be isomers if they have the same mass and charge but differ in state of excitation. The explanation of such isomerism was first given by Weiszäcker (1936). If the nucleus is excited to a state from which transitions to the ground state with emission of radiation are forbidden by selection rules, the nucleus may remain in the excited (isomeric) state for a time very long compared with the periods of internal nuclear motions. In practice an excited state of half-life greater than about 10^{-9} sec. is classified as an isomeric state. Weiszäcker envisaged the return of the isomeric nucleus to its ground state with the emission of γ-radiation. The transition rate for the emission of such radiation can be calculated in terms of a nuclear model and depends markedly on the change in angular momentum associated with the transition. It was found, however, that many nuclear isomers return to the ground state with the emission of internal conversion electrons. Evidently, then, the possibility of such transitions must be taken into account in estimating, from the lifetime, the angular momentum change associated with a transition. These radiationless transitions have in fact a marked influence on the transition rate in the case of highly forbidden transitions.

5.7. Detailed results of internal conversion coefficient calculations

We consider here the application of the general methods outlined in Chapter II, para. 2.6, to some particular cases. Until very recently the most extensive calculations of internal conversion coefficients had been carried out for the K and L shell conversion of heavy elements ($Z = 84$). Values of α_K were calculated for electric dipole radiation (Hulme, 1932), electric quadrupole (Taylor and Mott, 1932), magnetic dipole, quadrupole and octopole (Fisk and Taylor, 1934), while values of α_{L_1} were calculated for electric dipole and quadrupole radiation by Fisk (1934). The whole position with regard to these early calculations, which were made using an exact relativistic theory, were summed up by Hulme, Mott, F. Oppenheimer and Taylor (1936). More recent exact calculations of α_K and α_L for a number of heavy elements have been made by Stanley (1949) and by Gellman, Griffith and Stanley (1950). As before, these calculations were carried out for a considerable range of γ-ray energies. Stanley considerably simplified Hulme's formulae for the electric dipole case, but even so, the computational task involved was formidable. Recently, however, accurate and very extensive calculations of α_K and α_L for both electric and magnetic multipole radiation of energy above 150 keV. have been made by Rose, Goertzel, Spinrad, Harr and Strong (1949), (1951), using the Mark I Relay Calculator at Harvard. Some further calculations to investigate the effect of screening have been made by Reitz (1950). Earlier, non-relativistic calculations, valid for low-energy γ-rays and values of $Z < 40$ were made of α_K for electric multipole radiation by Hebb and Uhlenbeck (1938) and Dancoff and Morrison (1939), and were extended to include α_L by Hebb and Nelson (1940).

For magnetic multipole radiation a non-relativistic theory (which ignored electron spin) would not predict any internal conversion in the K shell or in any subshell with zero orbital angular momentum. This follows because a transition from an electron state with zero to one with l units of angular momentum (produced by conversion of 2^l-pole quantum radiation) involves a change of parity of the electron wave function if l is odd, but does not do so if l is even. On the other hand, magnetic 2^l-pole radiation can only produce a parity change if l is even and does not do so if l is odd. As pointed out by Dancoff and Morrison, this can be interpreted physically as arising from the fact that the magnetic multipole field cannot give rise to any radial forces on an s electron.

If spin is taken into account, however, transitions of electrons are possible under the influence of magnetic multipole radiation. Dancoff and Morrison (1939) calculated α_K approximately in this case, using a relativistic theory, but taking a plane wave to represent the ejected electron. Later, Goertzel and Lowen (1945) and Berestetzky (1947) independently tried to improve the theory by carrying out calculations which allowed for the distortion of the ejected electron wave, but they used a two-component Pauli theory of the spin. Later still, Drell (1949) showed this treatment to be unsatisfactory, since it omitted an important term, and himself carried out calculations using a Dirac relativistic theory and allowing for the distortion of the ejected electron wave. It may be noted that his formula reduces to that of Dancoff and Morrison in the limit of small Z, but he has assumed $v/c \ll 1$ throughout for the ejected electron. Similar calculations for the L-shell internal conversion of magnetic multipole radiation have recently been made by Lowen and Tralli (1949 a, b).

The following are some of the expressions which have been derived for the internal conversion coefficient in the K shell for the different types of radiation. $\alpha(l)$, $\beta(l)$ are the conversion coefficients for 2^l-pole electric and magnetic radiation respectively and ν is the quantum energy of the gamma radiation in units mc.^{2}†

(1) Electric multipole (non-relativistic):

$$\alpha_K(l) = \left(\frac{16}{137}\right)[\Gamma(l+\tfrac{1}{2})]^2 \left(\frac{l}{l+1}\right)\left(\frac{2}{\nu}\right)^{l+1}\frac{n^4}{(1+n^2)^{l-2}}$$

$$\times \frac{\{(l+1)(1+n^2)^{l-2}\exp(-2n\operatorname{arcot}n) - V_l\}^2}{(l^2+n^2)\{(l-1)^2+n^2\}\dots(1+n^2)\{1-\exp(-2\pi n)\}}, \quad (5.25)$$

---

* It is customary in the literature to use the symbol β to refer to internal conversion of magnetic multipole radiation. Thus β_K (l) is the i.c.c. for magnetic 2^l-pole radiation in the K shell. α_K (l) the i.c.c. for electric 2^l-pole radiation in the K shell. In presenting experimental results, however, the symbol α_K is used for the K shell i.c.c., irrespective of the multipole nature of the γ-radiation.

where

$$V_l = \frac{4^l \prod\limits_{j=1}^{l}(j^2+n^2)}{(1+n^2)^3(2l)!}\left[\frac{in(l+1)}{l+in}\,{}_2F_1\left(1,\ -2l;\ 1-l-in;\ \frac{1-in}{2}\right)-1\right]\Bigg\}$$

and
$$n = Ze^2/\hbar v = Z/137\beta,$$

$$(5.26)$$

v being the velocity of the ejected electron. If Z is small $(n\to 0)$, $v/c \ll 1$,

$$\alpha_K^{\text{el.}}(l) \fallingdotseq \{Z^3/(137)^4\}\,\{l/(l+1)\}\,(2/\nu)^{l+\frac{5}{2}}.\qquad(5.27)$$

If $n\to\infty$ $(v\to 0)$, $\alpha_K^{\text{el.}}(l)\to 0$.

(ii) Magnetic multipole (relativistic but $v_{\text{el.}}/c \ll 1$):

$$\beta_K(l) = \frac{l(l+2)}{(2l+1)(l+1)}\left(\frac{\nu^2}{4}\right)\{\alpha_K(l+1)\}$$

$$+\ \frac{2^{2l+4}\pi\,(l+1)}{137(Z/137)^{2l}(2l+1)\{\Gamma(l)\}^2}\cdot\frac{n^{2l+4}\prod\limits_{j=1}^{l-1}(j^2+n^2)}{(1+n^2)^{2l+1}\{1-\exp(-2\pi n)\}}.\qquad(5.28)$$

If Z is small $(n\ll 1)$,
$$\beta_K(l) \fallingdotseq \{Z^3/(137)^4\}(2/\nu)^{l+\frac{3}{2}}.\qquad(5.29)$$

For low-energy γ-rays, but $n\ll 1$,

$$\beta_K(l) \fallingdotseq \{(l+1)/4l\}\,(v/c)^2\alpha_K(l),\qquad(5.30)$$

while for high-energy γ-rays, but $n\ll 1$,

$$\beta_K(l) = \alpha_K(l) = 2Z^3/(137)^4\,\nu.\qquad(5.31)$$

If $n\to\infty$, $v\to 0$

$$\beta_K(l) = \frac{2^{2l+4}\pi\,(l+1)}{137\left\{\left(\dfrac{Z}{137}\right)^{2l}(2l+1)\,[\Gamma(l)]^2\right\}}.\qquad(5.32)$$

For the exact calculations the formulae are even more complicated, and the results are best represented graphically. This is done in para. 5.8, where the results of the calculations are compared with the experimental values.

5.8. The comparison of measured internal conversion coefficients with the calculated values

The calculations of Rose *et al.* (1949) on K-shell internal conversion coefficients have been carried out for twenty-three values of Z between 10 and 98, for fourteen values of the γ-ray energy between 150 keV. and 2·5 MeV. for each value of Z and for both electric and magnetic 2^l-pole radiation ($l = 1, ..., 5$). Similar calculations have been made for conversion in the three L shells.

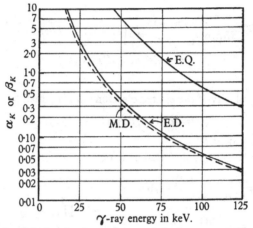

Fig. 28. Calculated values of α_K and β_K for copper ($Z = 29$).
E.D., E.Q. = electric dipole, quadrupole, etc.; M.D. = magnetic dipole.

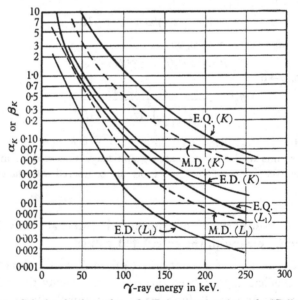

Fig. 29. Calculated values of α_K, β_K (Reitz, 1950) and α_{L_1}, β_{L_1} (Gellman, Griffith and Stanley, 1950) for indium ($Z = 49$).

Results of other exact calculations are shown in figs. 28–32. These figures show the I.C.C.'s for electric dipole, magnetic dipole and electric quadrupole radiations in the K shell of Cu(29) (fig. 28),

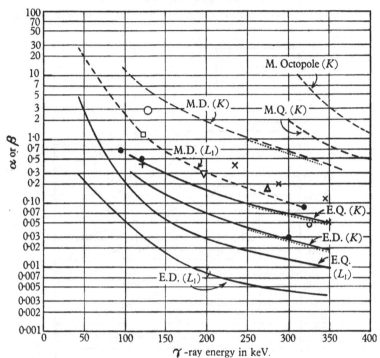

Fig. 30. Calculated values of α_K, β_K (Reitz, 1950) and α_{L_1}, β_{L_1} (Gellman, Griffith and Stanley, 1950) for E.D., E.Q. and M.D. conversion coefficients for polonium ($Z = 84$). Results for magnetic quadrupole and octopole radiation are also given (Fisk and Taylor, 1934). The calculations of Reitz allowed for screening. For comparison the dotted lines (·······) show the corresponding values of α_K and β_K calculated by Rose, Goertzel, Spinrad, Harr and Strong (1949) without allowing for screening. The points shown are experimental values for the total K shell internal conversion coefficient for elements near $Z = 84$ in atomic number. ∇ Ra (Stahel and Johner, 1934); ● Pt (Steffen, Huber and Humbel, 1949); × RaB, RaC (Ellis and Aston, 1930); □ Ta (Chu and Wiedenbeck, 1949) + Ir, △ Tl (Saxon, 1948); ○ Os (McCreary, 1942).

and in the K and L, shells of In(49) (fig. 29), Po(84) (fig. 30) and U(92) (fig. 31). The results shown for K-shell conversion were obtained by Reitz (1950), Hulme (1932), Griffith and Stanley (1949), and Hulme, Mott, F. Oppenheimer and Taylor (1936). In his calculations Reitz allowed for the screening of the atomic

nucleus by the orbital electrons by using electron-wave functions
for motion in a Fermi-Thomas field. In the other calculations the
effect of screening was neglected. The correction for screening is,
however, very small, as can be seen from fig. 30, which compares
Reitz's results with those given by other authors for Po(84).

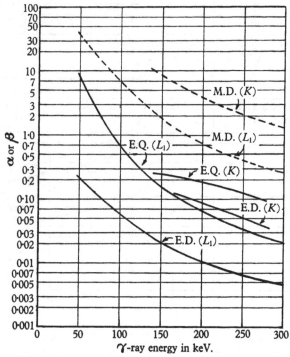

Fig. 31. Calculated values of α_K, β_K (Reitz, 1950) and α_{L_1}, β_{L_1} (Gellman,
Griffith and Stanley, 1950) for uranium ($Z = 92$).

Fig. 30 shows also the results of the exact calculations of Fisk and
Taylor (1934) of the K-shell internal conversion of magnetic
multipole radiation for Po(84). The exact calculations for internal
conversion in the L_{I} shell, the results of which are given in figs. 29,
30 and 31, were carried out by Gellman, Griffith and Stanley (1950).
Screening was neglected in these calculations also.

To indicate the way $\alpha_K(l)$ varies with multipole order, l, fig. 32
shows $\alpha_K(l)$ calculated by Segre and Helmholz (1949) using the
approximate expression (5.25) of Dancoff and Morrison (1939) for

$l = 1$ to 5. The curves given in this figure refer to $Z = 35$, but curves for other atomic numbers would exhibit similar features. Fig. 32 also shows similar curves for $\beta_K(l)$ ($l = 1$ to 4) calculated using (5.28) given by Drell (1949).

Results of calculations by Hebb and Nelson (1940) and by Lowen and Tralli (1949 a, b) for L-shell internal conversion are shown in fig. 33.

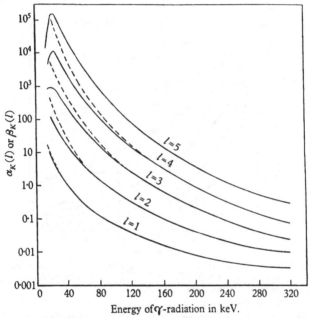

Fig. 32. Calculated values of $\alpha_K(l)$ ($l = 1$–5) and β_K (l) ($l = 1$–4) using approximate expressions (5.25) and (5.28). —— $\alpha_K(l)$; ---- $\beta_K(l)$.

The calculations illustrated by the results plotted in figs. 28–33 lead to the following general conclusions:

(1) Both $\alpha(l)$ and $\beta(l)$ increase rapidly with multipole order l for all values of Z and of the quantum energy of the γ-radiation.

(2) In general both $\alpha(l)$ and $\beta(l)$ increase rapidly with decrease of γ-ray energy, but for very low energies $\alpha(l)$ decreases to 0, while $\beta(l)$ continues to increase to a large finite value.

(3) Both $\alpha(l)$ and $\beta(l)$ increase with Z for a given γ-ray energy, the increase for small values of Z being nearly proportional to Z^3.

(4) The ratios $\alpha_K(l)/\alpha_L(l)$ and $\beta_K(l)/\beta_L(l)$ decrease with increase of multipole order l, and also with increase of atomic number Z, but they increase with increase of γ-ray energy for a given Z.

(5) Internal conversion is much more important in the L_I than in the L_{II} or L_{III} shells.

Measured values of α_K for γ-rays of a wide range of quantum energy from heavy elements are also given in fig. 30. The theoretical curves shown on this figure were calculated for $Z = 84$, and as the experimental points refer only to a range of Z between 73 and 88 they are fairly closely relevant to the curves. At higher energies many of

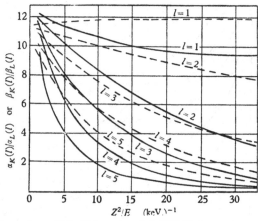

Fig. 33. The ratios $\alpha_K(l)/\alpha_L(l)$ (———) and $\beta_K(l)/\beta_L(l)$ (– – – –) as a function of Z^2/E ($E = \gamma$-ray energy in keV.) for $l = 1$–5. The calculations refer to $Z = 35$, but the curves are not very sensitive to Z.

the observed points fall close to either the electric dipole or electric quadrupole curves. At the lower energies, however, many of the γ-rays show internal conversion coefficients considerably in excess of those to be expected from the assumption of electric quadrupole radiation, approximating rather to the values expected for magnetic dipole radiation. These low-energy γ-rays were interpreted by Hulme, Mott, Oppenheimer and Taylor (1936) as due to transitions of the type $\Delta J = 1$, $\Delta L = 0$, i.e. to transitions between the various substates of a nuclear multiplet. Since there is no parity change in such transitions the selection rules (§ 5.3) indicate that the radiations would be mainly mixtures of magnetic dipole and electric quadrupole. Using their crude nuclear model described in § 5.4 it was calculated that the relative contribution to α_K from the magnetic dipole term compared with that from the electric quadrupole term should increase with the square of the wave-length for transitions of this type.* This trend is confirmed in fig. 30.

Support for this interpretation is provided also from the fact that the low-energy γ-rays come from odd nuclei that are known to possess a nuclear moment in the ground state. The even nuclei possess no spin angular momentum in the ground state, so that the γ-rays from them cannot be interpreted as due to inter-multiplet transitions.

A fairly direct test of the theory of internal conversion can be made in cases where the angular momentum change, ΔJ, in a nuclear transition can be assigned independently. For example, Brady and Deutsch (1950) have studied the angular correlation of the γ-rays of ^{60}Co (of energies 1·1715 and 1·3316 MeV. respectively) and found them both to be quadrupole and assigned to the spin of the three levels involved the values 0, 2 and 4. Angular correlation measurements, however, cannot distinguish between electric and magnetic multipole radiation, so that Brady and Deutsch were unable to determine the parity of the levels. Metzger and Deutsch (1950) have studied the correlation between polarization and direction of emission of these γ-rays and have considered that all three nuclear levels concerned must have the same parity.

Application of the rules given in Table XVIII shows therefore that both radiations must be electric quadrupole.

Waggoner, M. L. Moon and Roberts (1950) carried out an accurate measurement of the i.c.c.'s for these γ-rays using a thin lens, double-coil magnetic spectrometer of transmission 2·40% and line width 3·0% and applying equation (5.22) above. Thin sources were used, deposited on a thin three-ply laminated film of zapon-formvar-zapon. The counter had a thin mica window with a cut-off for about 50 keV. electrons.

Table XIX shows a comparison of the results for these radiations and also for the 560, 602, 799 keV. and 1·363 MeV. γ-rays of ^{134}Cs and the 1·11 MeV. γ-ray of ^{65}Zn with the calculated values of Rose *et al.* (1949). The table shows the transition type as predicted

---

* If g is the ratio of these two contributions the measured i.c.c. will be
$$\{g\beta_K(1) + \alpha_K(2)\}/(1 + g).$$

from angular correlation and polarization-directional correlation measurements.

For four of the radiations (1·17 and 1·33 MeV. of ^{60}Co, 602 and 799 keV. of ^{134}Cs) very good agreement is obtained. One value (the 1·11 MeV. radiation of ^{65}Zn) is slightly too high. In one case (the 560 keV. radiation of ^{134}Cs) the experimental accuracy is not good enough to permit an accurate comparison with theory, while in the other case (the 1·36 MeV. radiation of ^{134}Cs) there is disagreement with theory, but the decay scheme on which the experimental calculations are based is very uncertain. On the whole, then, the agreement between the measurements and the calculated values is very good.

TABLE XIX

Parent nucleus	Atom in which conversion occurs	γ-ray energy (MeV.)	$\alpha \times 10^4$ (exp.)	Classification of radiation†	Theoretical value of α ($\times 10^4$)
^{60}Co	^{60}Ni	1·1715	1·73 ± 0·06	E_2	$E_2 = 1·545$ $M_1 = 1·387$
^{60}Co	^{60}Ni	1·3316	1·29 ± 0·035	E_2	$E_2 = 1·175$ $M_1 = 1·034$
^{134}Cs	^{134}Ba	0·560	75 ± 18	E_2 and/or M_1	$E_1 = 22·5$ $E_2 = 61·7$ $M_1 = 97·5$
^{134}Cs	^{134}Ba	0·602	51·9 ± 3·5	E_2	$E_1 = 20$ $E_2 = 52·5$ $M_1 = 81·3$
^{134}Cs	^{134}Ba	0·799	24·8 ± 1·2	E_2	$E_1 = 10·6$ $E_2 = 26·3$ $M_1 = 41·2$
^{134}Cs	^{134}Ba	1·363	6·20 ± 0·34	?	$E_1 = 3·87$ $E_2 = 8·5$ $M_1 = 13·0$
^{65}Zn	^{65}Cu	1·114	2·3 ± 0·1	E_2 and/or M_1	$E_1 = 1·2$ $E_2 = 1·84$ $M_1 = 1·74$

† E_l signifies 2^l-pole electric while M_l signifies 2^l-pole magnetic.

Tables XX and XXI give the results of the measurement of the i.c.c.'s for a large number of γ-rays. The tables include tentative assignments of the radiation character in certain cases, based on the calculated values of α_K, α_L, β_K, β_L or their ratios.*

* If both magnetic and electric transitions can occur the ratio N_K/N_L of electrons ejected from the K and L shells respectively is

$$N_K/N_L = \{g\beta_K(l) + \alpha_K(l+1)\}/\{g\beta_L(l) + \alpha_L(l+1)\},$$

where g is as defined above.

5.9. Angular correlation between successive internal conversion electrons

When an atomic nucleus passes from an excited state B to the ground state A by way of a second excited state C, the directions of emission of the successive γ-rays or internal conversion electrons will be related to one another under certain conditions. A necessary condition for the occurrence of such a correlation is that the lifetime of the state C should be short compared with the period of precession of the nucleus in the external fields in which it is situated.

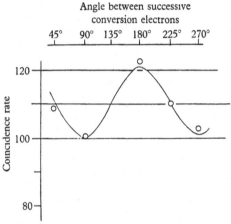

Fig. 34. Illustrating the angular correlation between successive conversion electrons ejected from the L shell of ^{197}Hg (Walter, Huber and Zünti, 1950).

The existence of such directional correlations between successive γ-ray quanta is now well established (Dunworth, 1940; Hamilton, 1940; Brady and Deutsch, 1947). Recently similar correlations have been observed by Frauenfelder, Walter and Zünti (1950) and by Walter, Huber and Zünti (1950) between the directions of successive internal conversion electrons ejected from the L shell of ^{197}Hg. Fig. 34 shows the curve obtained by these investigators for the number of coincidences in counters set to receive electrons ejected in directions making a variable angle θ with each other.

Calculations relative to this phenomenon have been made for a large number of types of conversion involving s electrons by Gardner (1949). Fig. 35 shows the calculated relative probability

TABLE XX

Parent nucleus	Atom in which conversion occurs	Half-life	γ-Ray energy (MeV.)	α_K	α_K/α_L	Remarks and other measurements	Assignment of multipole nature	References
$^{39}_{17}\text{Cl}$	$^{39}_{18}\text{A}$	55·5 m.	0·35	0·05	—	—	—	Haslam, Katz, Moody and Skarsgard (1950)
$^{44}_{21}\text{Sc}^{*}$	$^{44}_{21}\text{Sc}$	2·2 d.	0·269	0·08	8	—	E_4	Smith (1942)
$^{46}_{21}\text{Sc}$	$^{46}_{22}\text{Ti}$	85 d.	0·88	$1\cdot74\times10^{-4}$	—	—	E_2	Moon, Waggoner and Roberts (1950); Peacock and Wilkinson (1948)
			1·12	$0\cdot98\times10^{-4}$	—	—	E_2	
$^{51}_{24}\text{Cr}$	$^{51}_{23}\text{V}$	26·5 d.	1·0	0·001	—	—	—	Walke, Thompson and Holt (1940)
$^{52}_{25}\text{Mn}^{*}$	$^{52}_{25}\text{Mn}$	21 m.	0·392	5×10^{-4}	—	—	E_5+M_4	Osborne and Deutsch (1947)
$^{55}_{27}\text{Co}$	$^{55}_{26}\text{Fe}$	18·2 h.	0·477	$2\cdot5\times10^{-3}$	—	—	—	Deutsch and Hedgran (1949)
			0·935	$3\cdot8\times10^{-4}$	—	—	—	
			1·41	$1\cdot17\times10^{-4}$	—	—	—	
$^{58}_{27}\text{Co}$ $^{58}_{27}\text{Co}^{*}$	$^{58}_{26}\text{Fe}$ $^{58}_{27}\text{Co}$	72 d. 8·8 h.	0·805 0·0249	$2\cdot5\times10^{-4}$ —	1·9	—	E_2+M_1 E_4+M_3	Strauch (1950)
$^{60}_{27}\text{Co}^{*}$	$^{60}_{27}\text{Co}$	10·7 m.	0·059	—	4·5	$\alpha_L/\alpha_M=3\cdot1$	E_4	Deutsch, Elliott and Roberts (1945); Caldwell (1950)
$^{60}_{27}\text{Co}$	$^{60}_{28}\text{Ni}$	5·3 y.	1·1715	$1\cdot73\times10^{-4}$	—	—	E_2	Waggoner, Moon and Roberts (1950); Deutsch and K. Siegbahn (1950)
			1·3316	$1\cdot29\times10^{-4}$	—	—	E_2	
$^{61}_{29}\text{Cu}$	$^{61}_{28}\text{Ni}$	3·4 h.	0·076	Very large	10	—	E_2	Boehm, Blaser, Marmier and Preiswerk (1950)
			0·284	0·015	—	—	—	Owen, Cooke and Owen (1950)
			0·655	0	—	—	—	

Nuclide (studied)	Product	Half-life	E (MeV)	Rel. intensity	α_L/α_M	α	Assignment	References
$^{65}_{30}$Zn*	$^{65}_{29}$Cu	9.5 h.	0.0418	—	—	>6.4	E_1 or M_1	Hayward (1950a)
$^{65}_{30}$Zn	$^{65}_{29}$Cu	250 d.	1.114	2.3×10^{-4}	—	—	E_2 and/or M_1	Waggoner, Moon and Roberts (1950)
$^{69}_{30}$Zn*	$^{69}_{30}$Zn	13.8 h.	0.439	>0.01	—	—	E_5+M_4	Helmholz (1941)
$^{67}_{31}$Ga	$^{67}_{30}$Zn	78 h.	0.0925	<0.1	—	8	—	Valley and McCreary (1939); Alvarez (1938)
$^{81}_{34}$Se*	$^{81}_{34}$Se	57 m.	0.104	0.75	—	3.9	—	Bergström and Thulin (1949)
$^{80}_{35}$Br*	$^{80}_{35}$Br	4.54 h.	0.037	Very large	—	7	E_2	Berthelot (1944); West and Rothwell (1950); Siday (1941)
			0.049	1.2	—	7.3		
$^{79}_{36}$Kr	$^{79}_{35}$Br	1.30 d.	0.187	Very large	—	10	—	Creutz, Delsasso, Sutton, White and Barkas (1940)
$^{81}_{36}$Kr*	$^{81}_{36}$Kr	55 s.	0.127	—	—	4	—	Creutz, Delsasso, Sutton, White and Barkas (1940)
$^{83}_{36}$Kr*	$^{83}_{36}$Kr	—	0.032	—	—	0.44	E_4+M_3	Bergström, Thulin and Andersson (1950); Helmholz (1941)
		113 m.	0.046	—	—	1		
$^{85}_{36}$Kr*	$^{85}_{37}$Rb	4.5 h.	0.149	0.051	—	—	E_2+M_1	Bergström and Thulin (1950a)
$^{87}_{38}$Sr*	$^{87}_{38}$Sr	2.75 h.	0.390	0.25	—	6.9	E_5+M_4	Mann and Axel (1950); Helmholz (1941)
$^{87}_{39}$Y	$^{87}_{38}$Sr	80 h.	0.485	3.3×10^{-3}	—	—	E_2+M_1	Mann and Axel (1950)
$^{95}_{40}$Zr	$^{95}_{41}$Nb	35 d.	0.758	0.0024	—	—	—	Hudgens and Lyon (1949)†
$^{94}_{41}$Nb*	$^{94}_{41}$Nb	6.6 m.	0.0415	Very large	$\alpha_L/\alpha_M = 2.8$	0.31	E_4+M_3	Caldwell (1950); Goldhaber and Sturm (1946)
$^{93}_{42}$Mo*	$^{93}_{42}$Mo	6.75 h.	0.30	0.90	—	—	—	Kundu, Hult and Pool (1950)
			0.70	0.005	—	8.1		
$^{99}_{42}$Mo	$^{99}_{43}$Tc	67 d.	0.141	0.10	—	5	M_1	Medicus, Maeder and Schneider (1949); Bunker and Canada (1950)
			0.181	—	—	—	—	

† Medicus et al. (1950) attribute this line to $^{95}_{42}$Mo (see p. 130).

TABLE XX (continued)

Parent nucleus	Atom in which conversion occurs	Half-life	γ-Ray energy (MeV.)	α_K	α_K/α_L	Remarks and other measurements	Assignment of multipole nature	References
$^{94}_{43}$Tc	$^{94}_{42}$Mo	50m.	0·0344	—	1·2	—	E_4+M_3	Medicus, Preiswerk and Scherrer (1950)
			0·874	0·97×10⁻³	—	—	E_2+M_1	
$^{95}_{43}$Tc	$^{95}_{42}$Mo	60d.	0·201	0·036	7·1	—	—	Medicus, Preiswerk and Scherrer (1950);
			0·570	2·2×10⁻³	—	—	—	
			0·810	1·0×10⁻³	—	—	—	
		20h.	0·762	2·4×10⁻³	6·4	—	E_2	
$^{96}_{43}$Tc	$^{96}_{42}$Mo	104h.	0·312	—	—	—	—	Medicus, Preiswerk and Scherrer (1950)
			0·771 ⎫ 0·806 ⎬ 0·842 ⎭	5·9×10⁻⁴	—	mean α for three γ-radiations	E_1	
			1·119	2·7×10⁻⁴	—	—	E_1	
$^{97}_{43}$Tc*	$^{97}_{43}$Tc	90d.	0·096	—	1·6	—	E_5+M_4	Medicus, Preiswerk and Scherrer (1950); Helmholz (1941)
$^{104}_{45}$Rh*	$^{104}_{46}$Rh	4·3m.	0·080	0·6	—	—	E_2	Ageno (1943)
$^{106}_{45}$Rh	$^{106}_{46}$Pd	30s.	0·51	5·4×10⁻³	—	—	E_2	Metzger (1950)
			0·73	<2·5	—	—	E_2	
$^{106}_{47}$Ag	$^{106}_{46}$Pd	8·2d.	0·60	0·01	—	—	—	Feather and Dunworth (1938)
$^{107}_{47}$Ag*	$^{107}_{47}$Ag	44·3s.	0·094	16	0·92	$\dfrac{\alpha_K}{x_M+\alpha_N}=6\cdot0$	E_4+M_3	Bradt, Gugelot, Huber, Medicus, Preiswerk, Scherrer and Steffen (1946)

131

						$\dfrac{\alpha_K}{\alpha_M+\alpha_N}=6\cdot1$		
$^{109}_{47}$Ag*	$^{109}_{47}$Ag	40·5 s.	0·088	19	1·01	—	E_4+M_3	Bradt, Gugelot, Huber, Medicus, Preiswerk, Scherrer and Steffen (1946)
$^{110}_{47}$Ag*	$^{110}_{47}$Ag	60 d.	0·116	—	1·3	—	E_4	Siegbahn (1950)
$^{110}_{47}$Ag*	$^{110}_{48}$Cd	24·5 s.	0·656	$2\cdot5\times10^{-3}$	—	—	E_1 or M_1	
$^{111}_{47}$Ag	$^{111}_{48}$Cd	7·6 d.	0·340	0·015	—	—	E_2+M_1	Johansson (1950)
			0·243	<0·008	—	—	E_2+M_1	
$^{107}_{48}$Cd	$^{107}_{47}$Ag	6·7 h.	0·846	0·001	1·8	—	E_1	Bradt et al. (1946)
$^{111}_{48}$Cd*	$^{111}_{48}$Cd	48·7 m.	0·146	>11	>10	—	E_4	Hole (1948b)
			0·235	0·19		—	E_2	
$^{111}_{49}$In	$^{111}_{48}$Cd	2·7 d.	0·173	0·081	9	—	E_2+M_1	Boehm, Huber, Marmier, Preiswerk and Steffen (1949)
			0·247	0·036	6	—	E_2	
$^{113}_{49}$In*	$^{113}_{49}$In	104 m.	0·393	0·7	5·4	—	M_4	Lawson and Cork (1940)
$^{114}_{49}$In*	$^{114}_{49}$In	50 d.	0·192	~2·5	1	—	E_5+M_4	Boehm and Preiswerk (1949)
$^{115}_{49}$In*	$^{115}_{49}$In	4·53 h.	0·338	1	5·0	—	E_5+M_4	Lawson and Cork (1940)
$^{116}_{49}$In	$^{116}_{50}$Sn	54 m.	1·085	$8\cdot4\times10^{-4}$	—	—	E_2	Slätis, du Toit and Siegbahn (1950)
			1·274	$5\cdot7\times10^{-4}$	—	—	E_2	
$^{117}_{50}$Sn*	$^{117}_{50}$Sn	14 d.	0·157	Large	2·2	—	E_5+M_4	Mihelich and R. D. Hill (1950); Mallary and Pool (1950); Hayward (1950b)
			0·161	0·1	—	—	M_1	
$^{119}_{50}$Sn*	$^{119}_{50}$Sn	250 d.	0·069	—	1·5	—	M_4	Mihelich and R. D. Hill (1950)
$^{124}_{51}$Sb	$^{124}_{52}$Te	60 d.	0·607	$1\cdot6\times10^{-3}$	—	—	E_1	Langer, Moffatt and Price (1950)
$^{121}_{52}$Te*	$^{121}_{51}$Sb	17 d.	0·610	0·004	—	—	E_2	Katz, R. D. Hill and Goldhaber (1950)
$^{121}_{52}$Te	$^{121}_{52}$Te	143 d.	(0·082	—	0·75	—	E_5+M_4	R. D. Hill and Mihelich (1948)
			(0·213	0·085	7·3	—	E_2+M_1	

TABLE XX (continued)

Parent nucleus	Atom in which conversion occurs	Half-life	γ-Ray energy (MeV.)	α_K	α_K/α_L	Remarks and other measurements	Assignment of multipole nature	References
$^{129}_{52}$Te*	$^{129}_{52}$Te	90 d.	0.0885	—	0.68	—	E_5+M_4	Katz, R. D. Hill and Goldhaber (1950)
			0.159	0.18	8.7	—	E_2+M_1	
$^{125}_{52}$Te*	$^{125}_{52}$Te	58 d.	0.0354	>12	—	—	E_2+M_1	Bowe and Scharff-Goldhaber (1949); Friedlander, Perlman and Scharff-Goldhaber (1950)
			0.109	0.54	1.5	$\alpha_L/\alpha_M=3.5$	M_4	
$^{127}_{52}$Te*	$^{127}_{52}$Te	90 d.	0.0885	—	0.75	—	M_4	R. D. Hill (1949)
$^{129}_{52}$Te**	$^{129}_{52}$Te	32 d.	0.106	—	1	—	M_4	R. D. Hill (1949)
$^{131}_{52}$Te*	$^{131}_{52}$Te	29 h.	0.177	—	2	—	M_4	R. D. Hill (1949)
$^{130}_{53}$I	$^{130}_{54}$Xe	12.5 h.	0.537	6.9×10^{-3}	—	—	E_2	Roberts, Elliott, Downing, Peacock and Deutsch (1943)
			0.667	3.8×10^{-3}	—	—	E_2	
			0.744	3.1×10^{-3}	—	—	E_2	
			0.417	0.012	—	—	E_2	
$^{131}_{53}$I	$^{131}_{54}$Xe	8.0 d.	0.080	0.17	8.4	—	—	Kern, Mitchell and Zaffarano (1949)
			0.282	0.079	—	—	—	
			0.363	0.018	4.0	—	—	
$^{127}_{54}$Xe*	$^{127}_{54}$Xe	75 s.	0.175	—	5	—	E_4+M_3	Creutz, Delsasso, Sutton, White and Barkas (1940)
$^{131}_{54}$Xe*	$^{131}_{54}$Xe	12 d.	0.163	—	2.34	$\alpha_L/\alpha_M=3.4$	E_5+M_4	Bergström (1950)

Nuclide	Product	Half-life	Energy	Intensity	α	Ratios	Assignment	References
$^{133}_{54}$Xe*	$^{133}_{54}$Xe	5.27 d.	0.233	0.025	2.2	—	E_1	Bergström and Thulin (1950b)
	$^{133}_{55}$Cs	—	0.081	0.66	5.9	—	E_2+M_1	Peacock, Brosi and Bogard (1947)
$^{135}_{54}$Xe*	$^{135}_{54}$Xe	15.6 m.	0.52	0.2	—	—	E_5+M_4	Caldwell (1950); Wiedenbeck and Chu (1947); Peacock, Jones and Overman (1947)
$^{134}_{55}$Cs*	$^{134}_{55}$Cs	3 h.	0.128	0.0251	0.64	$\alpha_L/\alpha_M = 5.4$	—	
$^{134}_{55}$Cs	$^{134}_{56}$Ba	1.7 y.	0.560	7.5×10^{-3}			E_2 and/or M_1	Waggoner, Moon and Roberts (1950)
			0.602	5.2×10^{-3}			E_2	
			0.799	2.5×10^{-3}			E_2	
			1.363	6.2×10^{-4}			—	
$^{137}_{55}$Cs	$^{137}_{56}$Ba	33 y.	0.669	0.097	5.0	—	E_5+M_4	Osaba (1949); Waggoner (1950)
$^{133}_{56}$Ba*	$^{133}_{56}$Ba	38.8 h.	0.276	—	3.2	—	E_2	Cork and Smith (1941)
$^{141}_{58}$Ce	$^{141}_{59}$Pr	30.6 d.	0.145	0.254	5.5	—	E_2	Freedman and Engelkemeir (1950)
$^{153}_{62}$Sm	$^{153}_{63}$Eu	47 h.	0.1015	>0.7	5	—		J. M. Hill and Shepherd (1950)
$^{165}_{66}$Dy*	$^{165}_{66}$Dy	1.3 m.	0.109	—	0.126	$\alpha_L/\alpha_M = 3.75$; $\alpha_M/\alpha_N = 2.1$	E_4	Caldwell (1950); Hole (1948a)
$^{171}_{68}$Er	$^{171}_{69}$Tm	7.5 h.	0.113	1.3	—	—	E_2+M_1	Ketelle and Peacock (1948)
$^{170}_{69}$Tm	$^{170}_{70}$Yb	127 d.	0.0826	0.36	0.9	$\alpha_L/\alpha_M = 2.55$; $\alpha_L/\alpha_N = 3.3$	E_2	Grant and Richmond (1949); Saxon and Richards (1949); Fraser (1949)
$_{72}$Hf*	$_{73}$Ta	19 s.	0.150	>19	0.88	—	E_3	Hole (1948b)
$^{181}_{72}$Hf	$^{181}_{73}$Ta	46 d.	0.130	1.3	—	—	E_3+M_2	Fuller (1950); Lundby (1949); Jensen (1949); Chu and Wiedenbeck (1949)
			0.134	—	1.2		E_5+M_4	
			0.337	0.067	3.6		E_2+M_1	
			0.471	0.031	3.0		E_3+M_2	

TABLE XX (*continued*)

Parent nucleus	Atom in which conversion occurs	Half-life	γ-Ray energy (MeV.)	α_K	α_K/α_L	Remarks and other measurements	Assignment of multipole nature	References
$_{73}\text{Ta}^*$	$_{73}\text{Ta}$	16·5 m.	0·180	—	0·25	—	E_4	Hole (1948b)
$^{186}_{75}\text{Re}$	$^{186}_{76}\text{Os}$	91 h.	0·138	2·83	0·4	—	M_1	McCreary (1942)
$^{193}_{76}\text{Os}$	$^{193}_{77}\text{Ir}$	—	0·128	0·41	1·4	—	E_2	Saxon (unpublished), see Reitz (1950)
$^{192}_{77}\text{Ir}$	—	75 d.	—	0·32	—	—	—	Scherb and Mandeville (1948); Wiedenbeck and Chu (1947)
$^{194}_{77}\text{Ir}^*$	$^{194}_{77}\text{Ir}$	1·5 m.	0·0574	—	—	$\alpha_L : \alpha_M : \alpha_N$ $13\cdot2 : 3\cdot2 : 1$	E_4	Caldwell (1950)
$_{78}\text{Pt}^*$	$_{78}\text{Pt}$	3·45 d.	0·126	—	0·23	—	E_3	Hole (1948, b)
		78 m.	0·337	—	1·30	—	E_4	
$^{194}_{79}\text{Au}$	$^{194}_{78}\text{Pt}$	39 h.	0·291	0·034	2	—	E_1	Steffen, Huber and Humbel (1949)
			0·328	0·10	1·75	$\alpha_L/\alpha_M = 7\cdot5$	$E_2 + M_1$	
			1·48	0·0026	—	—	E_2	
$^{195}_{79}\text{Au}$	$^{195}_{78}\text{Pt}$	180 d.	0·096	2·5	4·7	$\alpha_L/\alpha_M = 6\cdot5$	$E_2 + M_1$	Steffen, Huber and Humbel (1949)
			0·129	1·0	5·5	$\alpha_L/\alpha_M = 5\cdot0$	$E_2 + M_1$	
$^{196}_{79}\text{Au}$	$^{196}_{78}\text{Pt}$	133 h.	0·358	0·050	1·7	—	E_2	Steffen, Huber and Humbel (1949)
$^{196}_{79}\text{Au}$	$^{196}_{80}\text{Hg}$	133 h.	0·175	0·10	3·3	—	E_1	Steffen, Huber and Humbel (1949)
$^{197}_{79}\text{Au}^*$	$^{197}_{79}\text{Au}$	7 s.	0·273	—	3·4	—	$E_4 + M_3$	Freuenfelder, Gugelot, Huber, Medicus, Preiswerk, Scherrer and Steffen (1948)
$^{198}_{79}\text{Au}$	$^{198}_{80}\text{Hg}$	2·76 d.	0·411	0·0280	2·25	$\alpha_L/\alpha_M = 4\cdot2$	E_2	Wiedenbeck and Chu (1947); Steffen, Huber and Humbel (1949)

Nuclide		Half-life	Energy			$N_{L_{III}}/N_{L_I}$	E_2+M_1	Reference
$^{199}_{79}$Au	—	3·3	0·1585	—	0·37	$=0.55$	—	R. D. Hill (1950)
			0·2085	—	~5		E_2+M_1	
$^{197}_{80}$Hg*	$^{197}_{80}$Hg	23 h.	{ 0·164 / 0·133	4·5	0·45	$\alpha_L/\alpha_{M+N}=2.23$	E_4 or E_5	Frauenfelder, Huber, De-Shalit and Zünti (1950)
	$^{197}_{79}$Au	23 h.	0·275	0·53	0·39	$\alpha_L/\alpha_{M+N}=2.9$	E_2	
	$^{197}_{79}$Au	65 h.	0·077	—	3·4	$\alpha_L/\alpha_{M+N}=3.6$	—	
$^{199}_{80}$Hg*	$^{199}_{80}$Hg	43 m.	0·362	>11	1·55	—	E_5+M_4	Hole (1948b)
^{80}Hg*	^{80}Hg	—	0·155	0·25	<0·43	—		Hole (1948b)
$^{203}_{81}$Tl	$^{203}_{81}$Tl	51·5 d.	0·286	0·18	3	$\alpha_L/\alpha_M>12$	E_2	Saxon (1948)
$^{204}_{82}$Pb*	$^{204}_{82}$Pb	68 m.	0·374	0·05	2·1	—	E_3	Sunyar, Alburger, Friedlander, Goldhaber and Scharff-Goldhaber (1950)
			0·905	0·10	1·5	—	E_6	
$^{238}_{93}$Np	$^{238}_{94}$Pu	2·1 d.	0·0422	—	—	$\alpha_L=0.71$, $\alpha_M=0.27$	—	Freedman, Jaffey and Wagner (1950)
			0·0468	—	—	$\alpha_L=0.74$, $\alpha_M=0.12$	—	
			0·1029	—	—	$\alpha_L=0.075$, $\alpha_M=0.047$	—	
			0·98	0·012	—	—	—	
			1·04	0·018	—	—	—	
$^{242}_{95}$Am	$^{242}_{96}$Cm	16 h.	0·052	—	—	$\alpha_L=0.5$	⟩	O'Kelley, Barton, Crane and Perlman (1950)

TABLE XXI*

Measured values of γ-ray internal conversion coefficients for naturally radioactive isotopes

Parent nucleus	Atom in which conversion occurs	γ-Ray energy (MeV.)	α_K	α_K/α_L	Other measurements and remarks	Assignment of multipole nature	References
Th C″ ($^{208}_{81}$Tl)	Th D ($^{208}_{82}$Pb)	0·0398	—	1·6	$\alpha_L = 0·8$; $\alpha_L/\alpha_M = 4$	—	Martin and Richardson (1950); Itoh and Watase (1941); Arnoult (1939); Flammersfeld (1939)
		0·210	—	1	—	—	
		0·232	—	3·5	—	—	
		0·250	0·25	10	—	—	
		0·277	0·07	—	—	$E_2 + M_1$	
		0·510	0·017	—	—	$E_2 + M_1$	
		0·582	0·017	—	—	E_2	
		0·859	0·0018	—	—	$E_2 + M_1$	
		2·62	—	—	—	$E_2 + M_1$	
Ra E ($^{210}_{83}$Bi)	Ra D ($^{210}_{82}$Pb)	0·047	—	—	$\alpha_L = 15·9$; $\alpha_M = 4·1$; $\alpha_N = 1·2$; $\alpha_0 = 0·3$	$E_2 + M_1$	Cranberg (1950)
Th C ($^{212}_{83}$Bi)	Th B ($^{212}_{82}$Pb)	0·114	—	—	$\alpha_L:\alpha_M:\alpha_N = 10·7:2·6:1$	—	Arnoult (1939); Martin and Richardson (1950); Itoh and Watase (1941)
		0·238	0·4	5·5	—	—	
		0·249	0·04	—	—	—	
		0·299	0·28	—	—	—	
Ra C ($^{214}_{83}$Bi)	Ra B ($^{214}_{82}$Pb)	0·053	—	—	$\alpha_L:\alpha_M:\alpha_N:\alpha_0 = 29:10·7:4·1:1$	—	Ellis and Aston (1930)
		0·241	0·36	8	$\alpha_L/\alpha_M = 4$	—	
		0·257	0·19	7	—	—	
		0·294	0·12	8	—	—	
		0·350	0·12	6·2	$\alpha_L:\alpha_M:\alpha_N = 5:2·5:1$	—	
Ac C ($^{211}_{83}$Bi)	Ac C″ ($^{209}_{81}$Tl)	0·350	0·10	—	—	E_2	Li (1937)

						E_2+M_1	
ThC ($^{212}_{83}$Bi)	ThC' ($^{212}_{84}$Po)	0·726	0·011	—	—	E_2+M_1	Martin and Richardson (1950); Itoh and Watase (1941)
		0·275	—	10	—	—	
		0·333	—	2	—	—	
		0·39	—	5	—	—	
		0·43	—	3	—	—	
RaC ($^{214}_{83}$Bi)	RaC' ($^{214}_{84}$Po)	0·607	0·0061	4·5	—	E_1	Rutherford, Chadwick and Ellis (1931); Ellis and Aston (1930); Itoh and Watase (1941); Constantinov and Latyshev (1941); Kulchitsky and Latyshev (1941); Ellis (1933b)
		0·766	0·0048	—	—	E_1	
		0·933	0·0061	5·9	—	E_2	
		1·120	0·0062	4·3	—	E_2	
		1·238	0·0057	—	—	E_2	
		1·379	0·0014	4·3	—	E_1	
		1·761	0·0016	4·5	—	E_2	
		2·193	0·0013	—	—	E_2	
An ($^{219}_{86}$Rn)	AcA ($^{215}_{84}$Po)	0·270	—	—	Mean $\alpha_{K+L}=0\cdot11$	—	Bennett (1938)
$^{226}_{88}$Ra	$^{222}_{86}$Rn	0·190	0·37	—	—	E_2+M_1	Stahel and Johner (1934)
UY ($^{231}_{90}$Th)	$^{231}_{91}$Pa	0·035	—	—	$\alpha_L=0\cdot82$	—	Knight and Macklin (1949)
UX$_1$ ($^{234}_{90}$Th)	UX$_2$ ($^{234}_{91}$Pa)	0·093	—	—	$\alpha_L=0\cdot34$ $\alpha_{M+N}=0\cdot07$	E_2	Bradt and Scherrer (1946)
UX$_2$ ($^{234}_{91}$Pa)	U$_{II}$ ($^{234}_{92}$U)	0·80	0·36	3·6	—	—	Bradt, Heine and Scherrer (1943); Bradt and Scherrer (1945)

The discrepancy between the measurements of different experimenters are large. The values of α must therefore be considered as subject to considerable doubt.

$I(\theta)$ of ejection of two successive internal conversion electrons in directions inclined at an angle θ to each other, for two possible nuclear transitions. The type of distribution is seen to depend markedly on the angular momenta of the initial and final nuclear states and on the orbital angular momenta of the ejected electrons. When further data on angular correlations have been accumulated they should provide important confirmatory evidence about the angular momenta of nuclear energy levels.

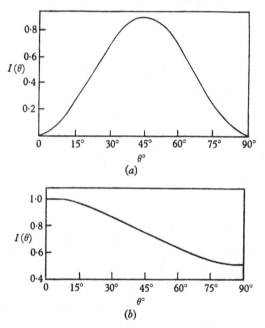

Fig. 35. Calculations by Gardner (1949) of the angular correlation between two successive internal conversion electrons. $I(\theta)\,d\theta$ is proportional to the probability that the angle between two successive conversion electrons should be between θ and $\theta+d\theta$. (a) $l_1 = l_2 = 0$; $J_A = 0$, $J_B = 2$, $J_C = 0$; (b) $l_1 = 3$, $l_2 = 2$; $J_A = 0$, $J_B = 3$, $J_C = 5$; where l_1, l_2 are the azimuthal quantum numbers of the ejected electrons and J_A, J_B, J_C the initial, intermediate and final total angular momentum quantum numbers of the nucleus.

5.10. Influence of internal conversion on the lifetimes of nuclear isomers

The lifetime of a nuclear isomer is proportional to the reciprocal of the total transition rate, p, from the isomeric state to a state of lower energy. In §5.4, equations (5.8)–(5.11) give expressions for p, calculated on the assumption of various nuclear models in turn. The isomeric half-value periods calculated using these expressions are those which would apply in the absence of extra-nuclear electrons. Owing to the presence of these electrons, however, the analysis of Taylor and Mott (§2.6) shows that the transition rate is increased, and the true total transition rate is nearly $p + b$, where b is the transition rate for internal conversion given approximately by the expressions (5.25)–(5.32) of §5.7. The half-value period τ of the isomeric state is then $\tau = 0.69/(p + b) = 0.69/p(1 + \alpha)$, in which α is the total internal conversion coefficient.

The magnitude of the theoretical half-value period depends on p, and through this on the assumed nuclear model. For a given l the calculated lifetime may differ by a factor of 10–100 for different models. However, for any given nuclear model, p depends markedly on the multipole nature of the radiation through a factor of the type $(Er_0/\hbar c)^{2l}$, where E is the energy of the emitted quantum and r_0 the nuclear radius.

TABLE XXII. *Influence of internal conversion on the lifetime of isomers*

Nucleus	Energy of transition (keV.)	Multipole order (l) of radiation	Half-life (observed) (sec.)	Half-life (calculated from 5.11) (sec.)	Half-life allowing for internal conversion (sec.)
$^{44}_{21}\text{Sc}$	288	5	$2\cdot1 \times 10^5$	$2\cdot6 \times 10^5$	$1\cdot67 \times 10^5$
$^{69}_{30}\text{Zn}$	430	5	$4\cdot97 \times 10^4$	5×10^4	5×10^4
$^{80}_{35}\text{Br}$	49	3	$1\cdot59 \times 10^4$	$1\cdot15 \times 10^6$	$0\cdot79 \times 10^4$
$^{87}_{38}\text{Sr}$	386	5	$9\cdot71 \times 10^3$	$1\cdot2 \times 10^4$	$1\cdot02 \times 10^4$
$^{107}_{47}\text{Ag}$	93·5	4	44·3	4×10^3	40
$^{113}_{49}\text{In}$	393	5	$6\cdot3 \times 10^3$	$1\cdot05 \times 10^4$	$6\cdot24 \times 10^3$
$^{114}_{49}\text{In}$	192	5	$4\cdot13 \times 10^6$	$8\cdot6 \times 10^6$	$4\cdot32 \times 10^6$
$^{115}_{49}\text{In}$	338	5	$1\cdot62 \times 10^4$	$2\cdot4 \times 10^4$	$1\cdot63 \times 10^4$

The quantity $Er_0/\hbar c$ is in general very small for the important isomeric transitions. For example, if E is 100 keV., $Er_0/\hbar c$ is approximately 5×10^{-3}. From this we conclude that for a given nuclear model the change in the calculated lifetime for a change of 1 unit in l is much greater than the change in the calculated lifetime from one model to another, for a fixed value of l.

The effect of internal conversion on the calculated half-value period may be very large for low-energy transitions. Table XXII shows half-lives calculated for some isomeric nuclei and for a number of different transition energies. In this table, which is based on one given by Berthelot (1944), the theoretical half-lives

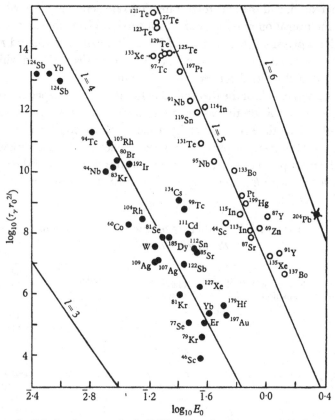

Fig. 36. Relation between the half-life and emitted γ-ray energy for nuclear isomers (Axel and Dancoff, 1949). The half-life for γ-emission is calculated as described in the text. The curves show the relationship to be expected on the basis of Bethe's nuclear model for different values of l.

shown are calculated using (5.11) of § 5.4 above, and also allowing for internal conversion, using the values of α calculated by Hebb and Uhlenbeck (1938) for electric multipole radiation. The comparison with the measured half-lives shows good agreement if the transition leading to internal conversion are allowed for.

The most detailed analysis of the information to be obtained from the measured half-lives concerning the angular momenta of isomeric nuclei has been made by Axel and Dancoff (1949).* Fig. 36 shows, for the known isomers of half-value period greater than 1 sec., the quantity $\log_{10}(\tau_\gamma r_0^{2l})$ plotted against $\log E_0$, where τ_γ is the estimated value of the half-life for each isotope neglecting internal conversion. τ_γ has been calculated from the expression $\tau_\gamma = \tau(1+\alpha)$, where τ is the measured half-life.

In making this calculation theoretical values of α were used because measured values of sufficient accuracy were not available. It was assumed that electric multipole transitions were involved, and a first approximation to the value of l was obtained by assuming $\tau_\gamma = \tau$ and using Bethe's formula (5.10) for τ_γ. The nuclei represented in fig. 36 fall into two groups. On the same figure curves calculated from (5.10) for $l = 3, 4, 5, 6$ are also plotted. The two groups of representative points cluster about the lines corresponding to $l = 4$ and $l = 5$, lending support to the interpretation of the half-lives given above. If no allowance had been made for internal conversion (i.e. if $\log_{10}(\tau r_0^{2l})$ had been plotted against $\log E$), some grouping of the points would still have been evident but it would not have been nearly so marked as in fig. 36.

Several review articles on nuclear isomerism are now available in English.†

5.11. Internal conversion following transitions of the type J = 0 → J = 0

In a few cases internal conversion electrons have been observed when the γ-ray from the corresponding nuclear transition could not be detected. Thibaud (1925) and Ellis and Aston (1930) drew attention to a case of this kind in Ra C′ which was later interpreted by Fowler (1930). The energy of the internal conversion electrons corresponded to that expected after a nuclear transition between two known states of the Ra C′ nucleus. But no γ-ray of energy 1·426 MeV. corresponding to this transition had been observed. Evidently the i.c.c. was very much greater than that of any other measured i.c.c. for a γ-radiation of comparable energy. Fowler

---

* See also Wiedenbeck (1946) and MacIntyre (1950).
† See, for example, Segre and Helmholz (1949), Hole (1948b), Devons (1949).

postulated that the two states involved in the transition each had a total nuclear angular momentum, $J = 0$, so that the radiative transition from one to the other was totally forbidden. Owing to the direct interaction between the nucleus and the atomic electrons, however, the nuclear transition could occur with the ejection of an electron from the atom. Transitions of this type are the exact parallel of the Auger transitions of the type $K \to L_I L_I$ observed in the magnetic spectra of electrons ejected by X-radiation (see §3.8).

A similar interpretation has more recently been placed on a group of internal conversion electrons ejected from ^{72}Ge (Bowe, Gold-haber, Hill, Meyerhof and Sala, 1948). These electrons correspond to a nuclear transition of energy $0·68$ MeV.* The γ-rays from this transition have not been detected, and the i.c.c. must be at least of the order of unity. Such a value of α, coupled with the high energy, would indicate an isomer of very long life, but the lifetime measured by Bowe et al. using delayed coincidences was $(5 \pm 0·5) \times 10^{-7}$ sec. This would seem to identify conclusively the transition as one of the type $J = 0 \to J = 0$.

Transitions $J = 0 \to J = 0$ can occur only between two states of the same parity. They are forbidden, even by the mode of internal conversion, if the states are of different parity. In this latter case transitions may occur through an intermediate state of different nuclear angular momentum. A mechanism for such transitions has been suggested by Sachs (1940). Suppose I and II (fig. 37) represent the ground state and the first excited state of a nucleus, and suppose the parity of the two states is opposite and $J = 0$ for each, so that transitions between them are completely forbidden.

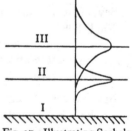

Fig. 37. Illustrating Sachs's mechanism for highly forbidden nuclear transitions.

If a third state III of different angular momentum, and of higher energy but of breadth sufficiently large for it to overlap state II, exists, transitions II → III and III → I should be possible, and should give rise to two quanta, two electrons, or one quantum and one electron. Transitions of this type would be distinguished by the emitted quanta and electrons having a continuous dis-

* Haynes (1948).

tribution of energy. Some evidence of such transitions has been produced by the observation of Goldhaber, Muehlhause and Turkel (1947) of an isomeric transition in [192]Ir (mean life 2·16 min., energy 58 keV.), which shows a continuous γ-ray spectrum as well as conversion electrons. Goldberger (1948) has shown that it would be possible to interpret this case as the ejection of a quantum and an electron by the mechanism suggested by Sachs.

Another example of a $J = 0 \rightarrow J = 0$ transition is exhibited by [80]Br (49 keV.).*

It is clear that the study of internal conversion in artificially radioactive isotopes is a rapidly developing subject of great interest in relation to nuclear models. There is obvious need for more accurate observational data in the field.

* Grinberg and Roussinow (1940).

CHAPTER VI

INTERNAL CONVERSION PROCESSES IN THE CREATION AND ANNIHILATION OF ELECTRON PAIRS

6.1. Internal pair production

γ-Radiation of quantum energy greater than $2mc^2$ may give rise to the process of the materialization of energy. The γ-ray quantum is absorbed and a pair of positive and negative electrons produced with total kinetic energy $h\nu - 2mc^2$, ν being the frequency of the radiation and m the rest mass of the electron. The process can occur only in the electrostatic field of a nucleus or of another electron, the extra particle being required to take off momentum and thus enable the conservation laws to be satisfied. Pair production may clearly occur in the neighbourhood of the nucleus in which the γ-ray quantum originates, in which case we have an internal conversion process with the production of electron pairs. Superficially this process appears to resemble closely that of the internal conversion of γ-rays with the ejection of bound electrons described in the previous chapter. In the new process, however, the interaction occurs between γ-rays and electrons in states of negative energy, which, according to the Dirac hole theory, are always present in the neighbourhood of nuclei.* The energy of the γ-ray is given up to one of these electrons and raises it to a state of positive kinetic energy, leaving a vacancy in the infinite distribution of electrons in states of negative kinetic energy. This vacancy behaves just like a particle of positive kinetic energy, positive mass and positive charge and is identified as a positron.

In spite of resemblances, there are, however, some important differences between the two processes of internal conversion. The interaction in the ordinary internal conversion process occurs between the γ-radiation and just a finite number of electrons. The process depends essentially on the acceleration of the electrons in the Coulomb field of the nucleus, and the internal conversion

---

* See, for instance, Spring (1950, Chap. v).

coefficient becomes very small for high-energy γ-rays and light nuclei. In contrast, however, for internal pair production, the nuclear γ-ray interacts with a large number of electrons in negative energy states. Also, the perturbation of the wave functions of these electrons by the field of the nucleus is rather small, so that internal pair production may occur even when the electrostatic field of the nucleus is negligible. This phenomenon is therefore not very sensitive to change in Z.

A new type of internal pair production has been observed recently in which, under the influence of the γ-radiation, an electron is raised from a negative energy state into a vacant inner shell. In this case the emitted positrons are homogeneous in energy.

6.2. Calculation of the coefficient of internal pair production

The coefficient of internal pair production, Γ, is defined as the ratio of the number of internal pairs produced to the number of γ-ray quanta emitted in nuclear transitions between specified states. The calculation of the internal pair transition rate follows lines similar to those for internal conversion and the relativistic Auger effect. In expression (2.9) of Chapter II one has to insert for ψ_i the wave function of an electron in one of the states of negative kinetic energy, and for ψ_f, as before, its wave function in a positive-energy state. Since in this case, however, we shall want to obtain expressions for the distribution in energy and angle of the electron pairs, it is convenient to write the transition rate for internal pairs in a slightly different form, viz.

$$b(\theta_-, \phi_-)\, d\Omega_-\, dt$$
$$= (2\pi k_- E_-/h^3 c^2) \left| \int \psi_f^*(\mathbf{r}_2)\{-ea_0 - e\rho_1 \mathbf{a}.\boldsymbol{\sigma}\}\psi_i(\mathbf{r}_2)\, d\mathbf{r}_2 \right|^2 d\Omega_-\, dt,$$
$$(6.1)^*$$

where $b(\theta_-, \phi_-)$ is the transition rate per unit time per unit solid angle for ejection of an electron into the direction of unit vector $\mathbf{n}_-(\theta_-, \phi_-)$. Expression (6.1) would give the transition probability if the initial state were discrete. However, the negative-energy states form a continuum and (6.1) has to be multiplied by a weighting factor $k_+ E_+\, dE_+\, d\Omega_+/2\pi c^2 h^2$. This represents the number of states

* In this and subsequent formulae the energy E is taken to include the mass energy.

in the negative-energy continuum with energy between $-E_+$ and $-(E_+ + dE_+)$, corresponding to motion in solid angle $d\Omega_+$ about the direction of unit vector $\mathbf{n}_+(\pi - \theta_+, \pi + \phi_+)$. $k_+\hbar$ is the magnitude of the corresponding momentum related to E_+ by

$$k_+^2 \hbar^2 = E_+^2/c^2 - m^2 c^2. \tag{6.2}$$

The expression then obtained for the rate of production of internal pairs in which the electron is moving (in solid angle $d\Omega_-$)

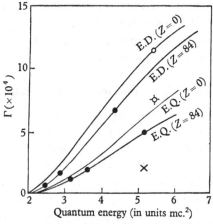

Fig. 38. Variation of coefficient of internal pair production, Γ, with energy of γ-radiation as calculated by Jaeger and Hulme for electric dipole and quadrupole radiations. The curves refer to calculations for a nucleus with $Z = 84$ and a hypothetical nucleus with $Z = 0$. The points shown refer to measured values of Γ. ● Points deduced from pair spectrum of ThC and RaC by Alichanow and his associates; ○ ^{24}Mg (2·78 MeV.) (Rae, 1949); ⟡ Mg (2·78 MeV.) (Mims, Halban and Wilson, 1950); × ThC″ (2·62 MeV.) (Bradt, Halter, Heine and Scherrer, 1946).

in direction $\mathbf{n}_-(\theta_-, \phi_-)$ with energy E_- and momentum $k_-\hbar$, while the positron is moving in solid angle $d\Omega_+$ in direction $\mathbf{n}_+(\theta_+, \phi_+)$ with energy between E_+ and $E_+ + dE_+$ and momentum $k_+\hbar$ is

$$b(E_+, \theta_-, \phi_-, \theta_+, \phi_+)\,d\Omega_-\,d\Omega_+\,dE_+ = (k_+ k_- E_+ E_-/h^5 c^4)$$
$$\times \left| \int \psi_f^*(k_- \mathbf{n}_-, \mathbf{r}_2)\{-e a_0 - e \rho_1 \mathbf{a}.\boldsymbol{\sigma}\} \psi_i(k_+ \mathbf{n}_+, \mathbf{r}_2)\,d\mathbf{r}_2 \right|^2 d\Omega_-\,d\Omega_+\,dE_+, \tag{6.3}$$

where ψ_i, ψ_f are normalized to have asymptotic form

$$\psi_i \sim \exp(-i k_+ \mathbf{n}_+.\mathbf{r}_2) + \exp(-i k_+ r_2) f_+(\theta, \phi)/r,$$
$$\psi_f \sim \exp(-i k_- \mathbf{n}_-.\mathbf{r}_2) + \exp(i k_- r_2) f_-(\theta, \phi)/r.$$

In (6.3) E_+, E_- are related by $E_- = h\nu - E_+$.

By integrating (6.3) over $d\Omega_-$, $d\Omega_+$ the energy distribution of the ejected positrons and electrons can be obtained. Alternatively, by integrating first over dE_+ the angular distributions of the positrons and electrons can be calculated. The total rate of internal pair production is obtained by carrying out all three integrations.

Equation (6.3) has been used to calculate b by Jaeger and Hulme (1935) for the case $Z = 84$ and for electric dipole and quadrupole transitions of quantum energy in the range $h\nu = 2$ to $7mc^2$. Similar calculations have also been carried out for the case $Z = 0$ where plane waves can be taken for ψ_i and ψ_f. This corresponds to Born's approximation. The calculations have been extended by Wang (1948) to the case of internal pair production by a magnetic dipole field. Fig. 38 shows the results obtained for Γ, the coefficient for internal pair formation. As pointed out in § 6.1, Γ varies very little with Z in marked contrast to the behaviour of the ordinary internal conversion coefficient. On the other hand, it should be remarked that external pair production (i.e pair production by plane electromagnetic waves) cannot occur for $Z = 0$. Physically, this is interpreted as arising from the fact that the conservation of momentum cannot then be satisfied.

Owing to the weak dependence of Γ on Z, most detailed calculations have been carried out using Born's approximation. Rose (1949) has given the following expression for $\gamma_l(E_+, \Theta)$, the number of pairs per unit energy interval of the positrons, per unit interval of $\cos \Theta$, per quantum, where Θ is the angle between the directions of ejection in a pair

$$\gamma_l(E_+, \Theta) = [2\alpha k_+ k_- m^2 c^3 (Kc/2\pi\nu)^{2l-1}/\pi\hbar K(l+1)\{(2\pi\nu)^2 - K^2 c^2\}]$$
$$\times [(2l+1)(E_+ E_- + m^2 c^4 - 3k_+ k_- c^2 \hbar^2 \cos \Theta)$$
$$+ l\{(Kc/2\pi\nu)^2 - 2\}(E_+ E_- - m^2 c^4 + k_+ k_- c^2 \hbar^2 \cos \Theta)$$
$$+ 3(l-1)k_+ k_- \hbar^2 c^2 \{(3/K^2)(k_- + k_+ \cos \Theta)(k_+ + k_- \cos \Theta)$$
$$- \cos \Theta\}] \qquad (6.4)$$

for electric 2^l-pole radiation, and

$$\gamma_l^m(E_+, \Theta) = [2\alpha k_+ k_- m^2 c^3 (Kc/2\pi\nu)^{2l-1}/\pi\hbar K\{(2\pi\nu)^2 - K^2 c^2\}]$$
$$\times [m^2 c^4 + E_+ E_- - (k_+ k_- \hbar^2 c^2 / K^2)(k_- + k_+ \cos \Theta)(k_+ + k_- \cos \Theta)]$$
$$\qquad (6.5)$$

for magnetic 2^l-pole radiation.

In these expressions $\mathbf{K} = \mathbf{k}_+ + \mathbf{k}_-$ gives the vector sum of the momenta of the electron and positron and α is the fine structure constant $(\doteqdot \frac{1}{137})$.

Integrating these expressions over Θ the following expressions are obtained for $\gamma_l(E_+)$, the number of positrons per unit energy range per quantum, ejected with energy E_+:

$$\gamma_l^{el.}(E_+) = (\alpha mc^2/\pi(l+1)\,h^3\nu^3)\,[\tfrac{1}{2}lh^2\nu^2 A_{l+1} + \{2lE_+E_- - \tfrac{1}{4}(2l+1)\,h^2\nu^2\}\,A_l$$
$$+ \{l(E_+^2 + E_-^2 + m^2c^4) + m^2c^4 - E_+E_-\}\,A_{l-1} - \tfrac{1}{4}(l-1)(E_+ - E_-)^2\,A_{l-2}]$$
$$(6.6)$$

for electric 2^l-pole radiation, and

$$\gamma_l^{m}(E_+) = (\alpha mc^2/\pi h^3\nu^3)\,[(E_+E_- + m^2c^4)\,A_l - \tfrac{1}{4}h^2\nu^2(A_{l+1} - x_1 x_2 A_{l-1})]\quad(6.7)$$

for magnetic 2^l-pole radiation.

In these expressions

$$A_l = \int_{x_1}^{x_2} x^l(1-x)^{-2}\,dx$$

and $\quad x_1 = (k_+ - k_-)^2\,c^2/4\pi^2\nu^2,\quad x_2 = (k_+ + k_-)^2\,c^2/4\pi^2\nu^2.$

The coefficient of internal pair formation, $\Gamma(l)$, is obtained by integrating the expressions (6.6) and (6.7) over all values of E_+ from mc^2 to $h\nu - mc^2$, and no closed expression has been given for the final result. However, for high-energy quanta, Rose and Uhlenbeck (1935) have given

$$\Gamma \doteqdot (2\alpha/3\pi)\{\log(2h\nu/mc^2) - \delta\},\quad\quad\quad(6.8)$$

where δ is $\tfrac{3}{2}$ and $\tfrac{61}{30}$ for electric dipole and quadrupole radiations, respectively.

Rose and Uhlenbeck (1935) have also given expressions valid approximately for light elements and γ-ray energies only slightly in excess of $2mc^2$, obtained using only those components of the Dirac four-component wave functions that have an analogy in the Schrödinger theory. These expressions are

$$\Gamma \doteqdot \tfrac{3}{4}\alpha^2 Zg^3\exp(-2\pi\alpha Z/g)\quad\quad\quad(6.9)$$

for electric dipole radiation and

$$\Gamma \doteqdot \tfrac{5}{12}\alpha^4 Z^3 g^3\exp(-2\pi\alpha Z/g)\quad\quad\quad(6.10)$$

for electric quadrupole radiation, where $g^2 = 2(f-2)$, fmc^2 being the quantum energy of the incident radiation.

6.3. Accuracy of the approximate calculations

To illustrate the conditions of validity of the approximate expressions for (6.8), (6.9) and (6.10) obtained by Rose and Uhlenbeck, fig. 39 shows Γ for different values of $h\nu/mc^2$ for electric dipole and quadrupole radiations. In each case curve E is the exact curve as calculated by Jaeger and Hulme for $Z = 84$, curve B that obtained using Born's approximation, and curve S that obtained using the 'Schrödinger' approximation (eqns. (6.9), (6.10)) that should be valid at low energies. For $h\nu/mc^2 = 6$ Born's approximation gives values 15 % too large for electric dipole radiation and 20 % too great for electric quadrupole radiation, and it becomes better at higher energies. On the other hand, the 'Schrödinger' approximation gives values somewhat too small.

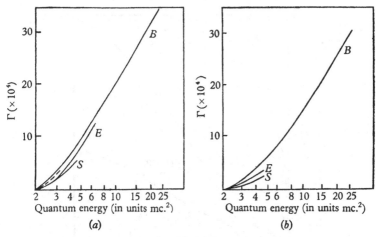

Fig. 39. Comparison of exact and approximate calculations of Γ for (a) electric dipole and (b) electric quadrupole transitions. E, exact theory; B, Born's approximation; S, Schrödinger approximation ($Z=84$). The broken curve in (a) is for the Schrödinger approximation with $Z=10$.

6.4. Energy distribution of the ejected particles

The energy distribution of the ejected positrons is not given very well by Born's approximation. This is illustrated in fig. 40, in which the broken lines show the distributions obtained by Born's approximation and the full lines the distributions calculated by Jaeger and Hulme for $Z = 84$, for electric dipole and quadrupole radiations of quantum energy $h\nu = 3mc^2$. It is seen that most of the energy is carried off by the positron, corresponding physically to its repulsion in the nuclear field. For higher energy γ-radiation the exact energy distributions of fig. 40 exhibit a maximum.

The mean distance, r, from the nucleus at which pair creation occurs is given approximately by

$$r \doteqdot 2Ze^2/(E_+ - E_-) = (Z/137)^2 \{2mc^2/(E_+ - E_-)\} a_K,$$

where a_K is the radius of the K shell. Thus, for the cases calculated by Jaeger and Hulme, the mean distance of internal pair creation is near the K shell radius. A correction should therefore have been made in their calculations for the effect of the K electrons in screening then uclear positive charge. This correction, however, should be small.

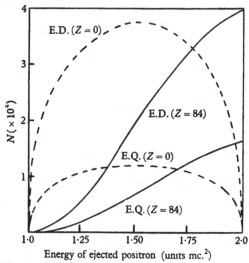

Fig. 40. Calculated energy distributions of conversion positrons for electric dipole and electric quadrupole radiations of quantum energy $h\nu = 3mc^2$ (1·536 MeV.) for $Z = 0$ and 84 respectively. N is the number of positrons ejected per unit energy range. —— exact calculations; ---- Born's approximation.

6.5. The coefficient of internal pair production

The variation of Γ with energy and multipole type of the radiation has been calculated by Rose (1949) and by Berestetzky and Shmush-kevich (1950). Figs. 41 (a) and (b) illustrate the calculations of Rose, using equations (6.4) and (6.5). They exhibit some important differences when compared with the curves for the ordinary internal conversion coefficient (figs. 28–32, Chapter v). Γ increases with increase of energy associated with the nuclear transition, decreases with increase of multipole order, and is greater for electric 2^l-pole than for magnetic 2^l-pole.

The increase with energy means that the internal pair production coefficient, Γ, becomes appreciable in just those ranges of γ-ray energy for which the internal conversion coefficient, α, is too small to be measured. This would suggest that measurements of Γ would

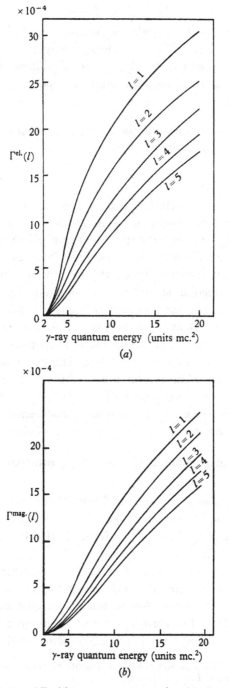

Fig. 41. Variation of Γ with quantum energy and multipole order l as calculated by Rose using Born's approximation. (a) Electric multipole radiations; (b) magnetic multipole radiations.

enable the assignment of angular momentum changes in transitions between nuclear levels differing widely in energy. The method has not so far been used extensively, and its usefulness may in fact be restricted because at higher energies Γ is not very sensitive to multipole order.

6.6. Angular correlation between directions of emission of electron and positron

The angular correlation between the positive and negative members of a pair has been calculated by Horton (1948) from equations (6.4), (6.5). He found it to depend on the multipole nature of the radiation concerned. Fig. 42 shows, as a function of quantum energy, the ratio R of the number of pairs per unit solid angle ejected with $\theta = 0$ and $\frac{1}{2}\pi$ respectively, where θ is the angle between the two particles. Evidently the experimental study of this angular correlation should provide an alternative method of determining the angular momentum change associated with a nuclear transition. Measurements of this kind have been made by Devons and Lindsay (1949) using the internal pairs emitted by ^{16}O formed in the reaction ^{19}F(p, α) ^{16}O, and they have found strong evidence to support the interpretation of the nuclear transition involved as one between two states of zero total angular momentum and even parity (see §6.8). Little other experimental evidence is available, however, on the angular correlations of electron pairs.

6.7. The experimental evidence about internal pair production

The most exhaustive experimental tests of the calculations of Γ and of the energy distribution of ejected positrons have been made by Alichanow and his co-workers* for internal pair production in Th$(C + C'')$ and Ra C.

In his measurements Alichanow used a transverse magnetic spectrograph and detected the positrons by Geiger counters used in coincidence. The active deposit was placed on a strip of aluminium, 10μ thick. The correction for positrons ejected from the aluminium by the γ-rays was determined experimentally and found

* Alichanow, Alichanian and Kosodaew (1936), Alichanow and Dzelepow (1938), Alichanow and Latyshev (1938).

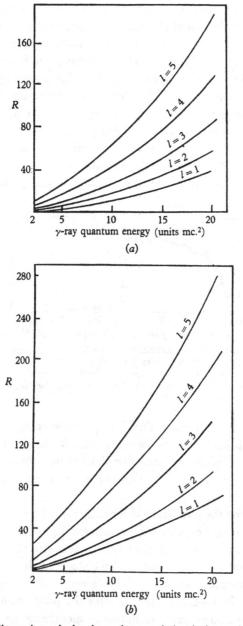

Fig. 42. Illustrating calculated angular correlation in internal pair formation. The curves show the ratio, R, of the number of internal pairs per unit solid angle ejected with $\theta = 0$ to the number with $\theta = \frac{1}{2}\pi$ for γ-radiation of different energy and multipole order ($l = 1$–5). (a) electric multipole radiations; (b) magnetic multipole radiations (Rose, 1949).

to be quite negligible from such a thin source. Fig. 40, which shows the energy distribution of the positrons calculated from the theory of Jaeger and Hulme, predicts a rise in intensity with increasing positron energy to a sharp cut-off at the maximum kinetic energy, $h\nu - 2mc^2$. Fig. 43, taken from an article by Latyshev (1947), gives the experimentally determined distributions obtained by Alichanow and Dzelepow for $Th(C + C'')$. The figure shows a number of sudden discontinuities, each of which is interpreted as the sharp cut-off corresponding to internal pair formation by one of the γ-rays of Th $(C + C'')$. The overall shape of the curve is determined by the form of the positron spectrum for internal pair production by

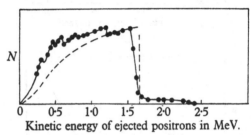

Kinetic energy of ejected positrons in MeV.

Fig. 43. Illustrating the measurements of Alichanow and co-workers of the energy distribution of internal pairs from $Th(C + C'')$. N is the number of positrons ejected per unit energy range. The broken curve shows the form of distribution expected for the 2·62 MeV. line from ThC'' (see fig. 40).

the well-known γ-ray from Th C″ of quantum energy 2·62 MeV. which is much more intense than any other in the spectrum. The theoretical form of the positron distribution from this conversion (shown broken in fig. 43) enables the experimental curve to be interpreted.

To determine the coefficient Γ for each γ-ray line the height of the corresponding break in the experimental positron curve was taken to give the number of positrons ejected with the maximum energy appropriate to each line, and the shape of the distribution in energy for that line was then taken as given by the Jaeger-Hulme calculations. The total transition rate for the production of γ-ray quanta of each energy was determined by measuring the spectrum of the recoil electrons ejected by the γ-radiation from a thick plate (see p. 115, footnote).

For the γ-ray line at 2·62 MeV. Alichanow (1938) obtained a value of Γ of 4–5 × 10^{-4}, in good agreement with the value 4·6 × 10^{-4} computed by Jaeger and Hulme for a quadrupole transition with this quantum energy. Fig. 38 indicates the experimental values of Γ obtained by Alichanow for various γ-radiations from Th (C + C″). The positions of the points with relation to the theoretical curves for electrical dipole and quadrupole radiations make it appear that some of the transitions are predominantly electrical dipole and some predominantly electrical quadrupole transitions.

A similar analysis has been made of the positron spectrum from Ra C. The ordinary internal conversion spectrum of this body has been extensively studied, and the conclusion has been drawn that all the transitions producing radiation sufficiently energetic to give rise to electron pairs are predominantly electric quadrupole in character. The estimated internal pair production coefficients, Γ, obtained by comparing the positron spectrum with the recoil electron spectrum, all lie on the electrical quadrupole curve.†

Further measurements of the coefficient of internal pair production have been made by Bradt, Halter, Heine and Scherrer (1946) for γ-radiation from Th C″ and by Rae (1949) and Mims, Halban and Wilson (1950) for the 2·78 MeV. radiation of ^{24}Mg emitted from a source of ^{24}Na. Rae used a magnetic spectrograph to measure both the positron energy distribution and the energy distribution of the β-radiation from a bare source of active sodium carbonate deposited on an aluminium foil 0·002 in. thick. The areas of the curves of the positron and electron distributions were used to obtain respectively the total rate of positron emission from the source and the total transition rate for the 2·78 MeV. transitions. The coefficient Γ was then determined by the ratio of these two rates.

Mims, Halban and Wilson measured the rate of emission of positrons by surrounding the ^{24}Na source with a plastic material and detecting the annihilation quanta by two naphthalene-anthracene crystals and photomultiplier tubes connected to record coincidences. The two counters were placed on opposite sides of the source at a distance of 65 cm. Annihilation coincidences could be distinguished by displacing the source out of the straight line joining

† This is consistent with the internal conversion results except for the line of quantum energy 1·379 MeV (see Table XXI).

the two detectors. The efficiency of the apparatus for positron detection was calibrated using standard γ-ray sources in lead or plastic containers. The effective positron strength of these sources could be calculated from data on pair production and γ-ray absorption.

To illustrate the difficulties encountered in using such measurements to assign the multipole nature of γ-radiation, Rae obtained for this 2·78 MeV. radiation of ^{24}Mg a value of Γ of $(11\cdot6 < 1) \times 10^{-4}$, and interpreted the radiation producing it as electric dipole. On the other hand, Mims, Halban and Wilson obtained the value $(8\cdot3 \pm 1\cdot1) \times 10^{-4}$ and interpreted the radiation as electric quadrupole.*

The conclusion emerges that the available experimental evidence can be interpreted satisfactorily in terms of the theory, but if internal pair production is to be used effectively for the assignment of the angular momenta of nuclear levels marked improvements in accuracy of the measurements is needed.

6.8. Internal pair production associated with transitions $J = 0 \rightarrow J = 0$

In a way precisely analogous to that discussed for internal conversion (Chapter v, §5.11), electron pairs can be produced as a result of the radiation-forbidden nuclear transition between two states of zero angular momentum and the same parity. A detailed study of the pairs produced in such a transition in the ^{16}O nucleus has been carried out by Rasmussen, Hornyak, C. C. Lauritsen and T. Lauritsen (1950). The maximum energy of the nuclear pairs in this transition is $5\cdot017 \pm 0\cdot03$ MeV. corresponding to an energy of $6\cdot04 \pm 0\cdot03$ MeV. for the upper $J = 0$ level concerned. The momentum distribution of the positrons produced in the transition is shown in fig. 44. The solid curve in this figure is the momentum distribution calculated by Oppenheimer (1941) on the assumption that the transition concerned is $J = 0 \rightarrow J = 0$. By contrast the broken curve shows the expected positron distribution for the internal conversion of electric dipole radiation. The energy dis-

* Rather indecisive evidence from γ-ray angular correlation measurements has been interpreted as indicating the radiation to be probably quadrupole (Brady and Deutsch, 1950).

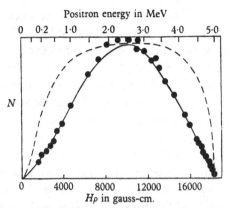

Fig. 44. Momentum distribution of the positrons ejected in the 6·0 MeV. transition in ^{16}O. N is the number of positrons per unit momentum range. The experimental points are those measured by Rasmussen, Hornyak, C. C. Lauritsen and T. Lauritsen (1950). The full curve is that calculated by Oppenheimer (1941) on the assumption of a $J = 0 \rightarrow J = 0$ transition. The broken curve is the corresponding distribution for internal pair production by electric dipole radiation.

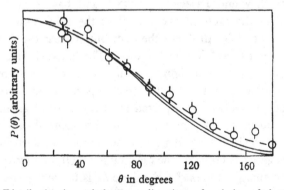

Fig. 45. Distribution in angle between directions of emission of electrons and positrons emitted from the 6·0 MeV. pair emitting transition of ^{16}O. $P(\theta)d\theta$ is the number of pairs in which the angle between electron and positron lies between θ and $\theta + d\theta$. The experimental points shown were obtained by Devons and Lindsay. The experimental curve (----) is fitted by the expression $P(\theta) = 1 + 0.85 \cos \theta$. The theoretical curves (——) refer to ideal geometry (lower curve) and to actual geometry with finite slit width (upper full curve).

tribution for positrons produced by higher-order multipole radiation is even flatter. The excellent agreement between the calculated and observed distributions is convincing evidence of the correctness of the interpretation of the pairs as arising from a $(0 \rightarrow 0)$ transition.

Devons and Lindsay (1949) have measured the angular correlation between the direction of ejection of the positrons and electrons produced in the same transition. Fig. 45 shows a comparison between their observations and the theoretical curve calculated by Oppenheimer (1941) for a $J = 0 \rightarrow J = 0$, even-even transition. The agreement is good over most of the angular range, but the measurements show rather more pairs ejected at large angles than are predicted by the theory.* Estimates of the lifetime of the excited $J = 0$ scale of ^{16}O have been made by Devons, Hereward and Lindsay (1949).

6.9. Homogeneous positrons from γ-ray emitters

A careful study of the energy distribution in the positron internal conversion spectrum of Ra C′ has been made recently by Gei, Latyshev and Pasechnik (1948)† and has revealed a fine structure in the high energy cut-off of the continuous positron spectrum. Thus fig. 46 shows the distribution near the cut-off for the 1·527 and 1·620 MeV. lines‡ of Ra C′. The statistical error in each of the points given in this figure amounts to about ± 8 in the recorded counts of about 650, so that the effect appears to be established.

The effect has been interpreted by Sliv (1949) as arising from internal conversion of γ-radiation from a nucleus already ionized in an inner shell. The γ-radiation is used to raise an electron from a state of negative energy into the vacant inner shell, leaving a positron of kinetic energy $h\nu - 2mc^2 + E_s$ where $h\nu$ is the γ-ray energy and E_s the ionization energy of the vacant inner shell. The kinetic energy of the positrons produced in this process is thus greater than the energy of cut-off $(h\nu - 2mc^2)$ for the positrons in the continuous

---

* Thomas (1940) has calculated the transition rate for the production of internal pairs in the 1·426 MeV. $(J = 0 \rightarrow J = 0)$ transition of Ra C′ (see §5.11). The transition rate is small, however, and it is not surprising that the pairs have not been observed from it.

† See also Latyshev, Gei, Bashilov and Barchuk (1948).

‡ These lines, observed in the recoil electron spectrum by Latyshev (1947), have not been observed in the internal conversion spectrum.

spectrum. Latyshev and co-workers (1948) have succeeded in identifying peaks corresponding to the filling of the K, L and M shells in this way. Thus in fig. 46 the line marked L 1527 occurs at an energy of 17 keV. above the cut-off for the 1·527 MeV. radiation of RaC′, that marked K 1527, occurs 95 keV. above and that marked M 1620 occurs 4 keV. above the cut-off of the 1·620 MeV. line. The K, L and M ionization energies of Ra C′ are very close to 95, 17 and 4 keV. respectively.

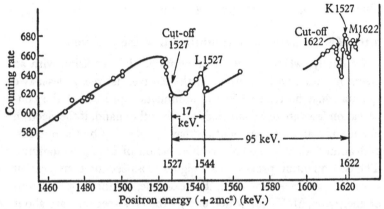

Fig. 46. Peaks of mono-energetic positrons produced by the capture of electrons of negative total energy into vacant inner shells following internal conversion of γ-radiation from Ra C′.

The probability of producing homogeneous positron groups may be quite appreciable in the case where a nucleus emits a complicated γ-ray spectrum in which one of the first radiations emitted has a low energy and thence a relatively high internal conversion coefficient while a subsequent radiation has a quantum energy exceeding $2mc^2$.

Let Γ_γ, Γ_s be respectively the widths of the excited level in the nucleus associated with the γ-ray emission, and the total width of the s inner level. If α_s is the total internal conversion coefficient of the atom by all the γ-rays emitted prior to the process, the production rate of s ionized atoms is $\hbar\Gamma_\gamma/\alpha_s(1+\alpha_s)$. Let b_μ^s be the transition rate for filling of the s shell of an s ionized atom by the process considered, then the actual transition rate will be

$$b_m^s = b_\mu^s \alpha_s \Gamma_\gamma / (1 + \alpha_s) \Gamma_s.$$

If b_p is the total transition rate for the emission of positrons into the continuous spectrum, the quantity b_m^s/b_p can be measured and is given by

$$b_m^s/b_p = (b_\mu^s/b_p)\, \alpha_s \Gamma_\gamma/(1+\alpha_s)\, \Gamma_s. \qquad (6.11)$$

Sliv calculated b_μ^s/b_p for a number of different multipole types of γ-radiation and from the measured values of b_m^s/b_p and the known values of Γ_s it was possible to calculate Γ_γ. For the peak K 1527 due to the conversion of the 1·527 MeV. radiation leading to the filling of the K shell, Latyshev and co-workers obtained $b_m^s/b_p = 3 \times 10^{-3}$ leading to the value of 0·4 eV. for Γ_γ.*

6.10. The radiationless annihilation of the positron

Generally, when a positron is annihilated in collision with an electron, radiation is emitted. If the electron is free, at least two quanta must be emitted in the annihilation process, if the conservation laws are to be satisfied. On the other hand, if the electron is bound in an atom, momentum may be taken up by the nucleus, and annihilation is possible with emission of a single quantum. The two-quantum process has a higher cross-section than has that of one-quantum annihilation. For example, for incident positrons of energy 0·5 MeV., for which energy the cross-sections are about at a maximum, the cross-section for the single-quantum process in lead is $1·93 \times 10^{-23}$ cm.^{2} compared with $3·8 \times 10^{-22}$ cm.^{2} for the two-quantum process.

On the hole theory the one-quantum process is interpreted as arising from a radiative transition in which an electron bound in an atom falls into a vacant level in the continuous distribution of negative-energy states. Owing to the presence of other electrons in the atom it is clearly possible for the excess energy liberated in such a transition to be used in the ejection of another bound electron from the atom, the annihilation thus taking place without radiation. The process is clearly closely analogous to the Auger effect and the method of calculation is similar.

If E_1, E_2 are the energies of the electrons 1 and 2 in the atom, and E_+, E_- those of the incident positron and ejected electron,

$$E_+ + E_1 = E_- - E_2.$$

---

* This assumes α_s to be of the order of unity which would be the case if 1·527 MeV radiation were preceded by a low energy transition.

Calculations of the cross-section for this process have been made by Massey and Burhop (1938). In these calculations plane waves were used for the ejected electron, and allowance was made for the distortion of the positron-wave function by the Coulomb field of the nucleus. Table XXIII shows the calculated cross-sections for the annihilation of a positron by one of the K electrons of lead, the excess energy being used to eject the other K electron. The table also illustrates the effect of the repulsion of the positron by the nucleus in reducing the cross-section.

TABLE XXIII. *Partial cross-sections (in units of 10^{-26} cm.²) for radiationless annihilation of positrons in lead, involving the two K electrons*

Energy of incident positrons in eV.	10^5	3×10^5	5×10^5
Cross-section allowing for nuclear repulsion	0·1	0·35	0·2
Cross-section assuming plane wave for incident positron	1·6	1·4	0·9

Approximate calculations for the relative annihilation cross-sections by other pairs of electrons gave the ratio

$$KK:KL:KM:LL = 1\cdot0:1\cdot0:0\cdot3:0\cdot25.$$

It appears, therefore, that the total cross-section for radiationless annihilation in lead has a maximum value of about 10^{-26} cm.² for incident positrons of energy 3×10^5 eV. The corresponding value for radiative one-quantum annihilation is $1\cdot9 \times 10^{-23}$ cm.², that is, about 2000 times as large. The two-quantum cross-section is twenty times larger again, thus for all positrons of this energy annihilated in lead, about 1 in 40,000 should give rise to the radiationless ejection of a fast electron.

The effect has not yet been observed because of its rather low probability. It should lead to a characteristic appearance, however, a slow positron making a collision with a lead atom and an electron of much great kinetic energy being ejected. The radiationless annihilation probability will increase with Z, approximately proportionally to Z^2.

CHAPTER VII

THE AUGER EFFECT IN MESON CAPTURE

The Auger effect plays an important role in the phenomenon of the capture of mesons by nuclei. Slow electrons have been observed at the ends of μ-meson tracks by Cosyns, Dilworth, Occhialini, Schoenberg and Page (1949),* and also accompanying stars associated with the capture of π-mesons by Menon, Muirhead and Rochat (1950). These slow electrons have been interpreted as Auger electrons emitted following meson capture by atoms in a photographic emulsion. Plates IV and V illustrate typical events of this type accompanying μ- and π-meson capture respectively. Evidence has also been obtained by Chang (1949) of the emission of radiation in the transition processes following μ-meson capture. He found a number of cloud-chamber tracks of Compton electrons which, from their directions of motion, appeared to have been produced by γ-radiation originating at the ends of μ-meson tracks.

The possibility of Auger processes involving mesons is of importance in the interpretation of the lifetime of negative mesons. A μ-meson may decay with the emission of a fast electron, together with one or more neutral particles. From cosmic-ray observations the lifetime for this process is known to be 2×10^{-6} sec. This is always the observed lifetime of positive μ-mesons which, owing to their charge, cannot be captured by an atomic system. For negative μ-mesons, however, the observed lifetime depends on the atomic number of the absorbing material. For absorption by light elements it is almost the same as for positive μ-mesons. For medium or heavy elements the negative μ-meson lifetime decreases with increase of Z according to a relation of the type

$$1/\tau = 1/\tau_0 + 1/\tau', \tag{7.1}$$

in which τ_0 is the positive μ-meson lifetime and τ' approximately proportional to Z^{-4}, is evidently the lifetime for the competing process of loss of the negative μ-meson. This competing process

---

* See also Fry (1950) and Franzinetti (1950).

PLATE IV

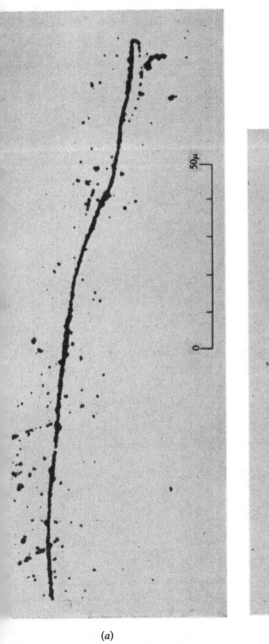

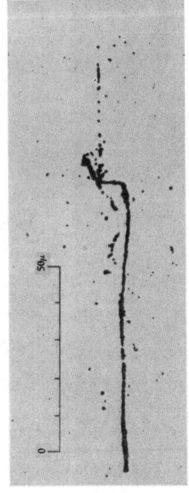

(a)

(b)

For explanation see p. xii

is now believed to be the interaction with a proton in the nucleus to form a neutron and a neutrino (ν), i.e.

$$\mu + p \rightarrow n + \nu. \tag{7.2}$$

The life history of the negative μ-meson leading up to its eventual decay into an electron or capture by a proton inside a nucleus is believed to be somewhat as follows: The meson produced originally with an energy of many millions of electron-volts loses energy in its passage through an absorbing material in inelastic collisions with the absorber atoms or molecules. So long as the velocity of the meson is greater than the velocity of the electrons in the atoms of the absorber, the ordinary considerations concerning stopping power of charged particles apply.

In the case of a gaseous absorber, when the meson has been slowed down to an energy of a few electron-volts it may be captured by a gas molecule A in a process $A + \mu^- \rightarrow A^* + e$, where A^* refers to a neutral molecule that has had one of its orbital electrons replaced by a μ-meson and e is the ejected electron. The mechanism by means of which this process can occur has been considered in some detail by Wightman (1950) for the case of capture by atomic hydrogen. If a negative μ-meson moving with a velocity small compared with the velocity of the electron in an hydrogen atom approaches the nucleus of the atom, the electron has to be regarded as moving in the field of an electric dipole. But there are no stable energy levels for an electron in such a field, so that it is ejected from the hydrogen atom and replaced by a meson.

A similar process should also be possible for capture of a negative meson by a heavier atom or by a molecule, but it has not so far been discussed in detail. Alternatively, a slow μ-meson might be captured by an atom or molecule in a process of radiative attachment to form a type of negative ion. The relative probability of these two types of process of meson capture by an atomic system is difficult to assess. In either case a μ-meson is associated intimately with an atomic system. Since the meson carries a negative charge equal to that of an electron, stable states of motion for it will exist in the field of the nucleus and the other electrons.

If the absorbing material is a metal the processes by which a μ^--meson eventually becomes attached to a single atom are somewhat different. The meson loses energy in collisions with the free conduction electrons of the metal. Eventually, however, the meson will be moving in an orbit of radius about equal to that of an inner shell, and thereafter the processes to be considered are the same whether the absorber is gaseous or solid. The case of a metallic absorbing material has been discussed by Fermi and Teller (1947).

If the field in which the meson moves can be taken as approximately hydrogenic the energy of the state of total quantum number n will be

$$-(\mu c^2/2)(Z/137)^2/n^2,$$

and the radius of the corresponding Bohr orbit, $a_n^{(\mu)}$, will be given by

$$a_n^{(\mu)} = n^2\hbar^2/Ze^2\mu,$$

where μ is the meson mass.

The radius of a circular meson orbit is much less than that of the circular electron orbit with the same value of n owing to the much greater mass of the meson. For example, the total quantum number n for the meson in a circular orbit of radius equal to that of an electron K orbit is $n = (\mu/m)^{\frac{1}{2}}$. The meson binding energy in such an orbit is equal to the K-electron ionization energy.

In general, it might be expected that the meson will first be captured into a state of high excitation whose wave function will have little overlap with the nucleus. Before it can interact appreciably with one of the nuclear protons it must drop into a state of low excitation with considerable overlap with the nucleus. In this process of 'meson de-excitation' both radiative and Auger transitions can occur, the latter with ejection of further electrons. If the meson has been initially captured by a molecule, it seems likely that, as the transition process goes on, sufficient energy will be transferred to the nuclear motion to cause molecular dissociation, so that thereafter it is sufficient to consider the meson as moving in an atomic field.

Owing to the small radius of the meson orbits of low excitation (say $1s$, $2s$, $2p$ orbits) in a heavy nucleus, a meson in such orbits may spend a considerable fraction of its time inside the nucleus, so that the reaction (7.2) may occur with appreciable probability. For these states it is not legitimate to take the field in which the meson moves as hydrogenic. Wheeler (1949) has calculated the position of the meson energy levels in elements of different Z by taking the nucleus to be equivalent to a uniform charged sphere of radius $R = e^2A^{\frac{1}{3}}/2mc^2$ (A = atomic mass number), so that the field in which the meson moves is given by

$$
\begin{aligned}
V(r) &= -(Ze^2/R)\{\tfrac{3}{2}-r^2/2R^2\} \quad &(r<R) \\
&= -Ze^2/r \quad &(r>R).
\end{aligned}
\qquad (7.3)
$$

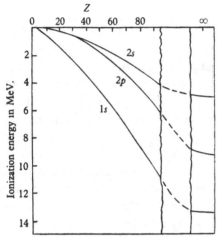

Fig. 47. Variation with Z of the ionization energy (in MeV.) for μ-mesons captured in $1s$, $2s$ and $2p$ meson levels (Wheeler, 1949).

If the nucleus were very heavy the motion of the meson would be confined entirely to the region $r<R$. In this region the form of potential is that of a harmonic oscillator (force directly proportional to distance). It is a characteristic of such a central force that the $2s$ level lies higher than the $2p$ level. Even in actual nuclei, where only part of the motion in the $2s$ and $2p$ orbits takes place inside the nucleus, the $2s$ level lies higher than the $2p$.

Fig. 47 shows the energy in MeV. of the $1s$, $2s$ and $2p$ states as a function of Z for a μ-meson as calculated by Wheeler using the potential (7.3). In the limit of very large Z the energy levels go over to those characteristic of a harmonic oscillator potential.

It is of great interest to inquire whether the lifetime τ' of (7.1) is determined primarily by the time (t_1 say) required for the meson to be slowed down until it is captured by an atom or molecule, by the time (t_2 say) taken to drop down from the orbit of high total quantum number in which it is captured to a state of low excitation, or by the time (t_3 say) spent in a $1s$ state before interacting with a nuclear proton.

The time t_1 depends markedly on the material in which the meson is slowed down. It could clearly be comparatively long in a gas, such as hydrogen, of low stopping power, but much shorter for a solid material of high stopping power. For example, Wightman (1950) estimated that μ-mesons in H_2 gas at a pressure of 1 atm. would require about 7×10^{-7} sec. to be reduced in energy from 10 MeV. to 0·5 eV., while the corresponding time for liquid hydrogen at $-253°$ C. would be about 9×10^{-10} sec.

The most extensive calculations of the time t_2 of transition to a state of low excitation after capture have been made by Burbidge (1951) and de Borde.† They considered in detail the transition rate for a μ^--meson captured in a circular orbit in the neighbourhood of the electron K shell and dropping down in steps to the meson orbit through circular orbits of total quantum number decreasing by 1 in each step. They found that for light elements Auger transitions were much more probable than radiative transitions when they were just energetically possible, but radiative transitions became more important for transitions between states of very low excitation. Wheeler (1949) showed that for $Z > 25$ the transition rate for the production of internal electron pairs becomes appreciable for the final transition for the $n = 2$ to $n = 1$ orbits. Table XXIV shows a selection of the results of the calculations. It is clear from the table that even for light elements the high probability of Auger transitions ensures that the time required to drop to the $1s$ state from an orbit in the neighbourhood of the K-electron orbit is not greater than 10^{-11} sec., while for values of $Z > 15$ it is considerably less. As already mentioned, the calculations of Auger transition rates concern only circular orbits. Some results for radiative transition between orbits for which $l < n - 1$ are shown in Table XXIV, and these are sometimes considerably smaller than for circular orbits. It is not possible, however, without further calculations, to estimate the total time of transition to the $1s$ orbit in these cases, although it seems unlikely it would be

† Unpublished.

longer by several orders of magnitude than the time calculated for circular orbits.

Since the Auger transitions play such an important role in shortening the time required for the meson to reach its lowest orbit, it is important to consider whether they could be inhibited in any way. Each Auger transition requires the ejection of an inner electron from the atom. In the case of Auger transitions involving ejection of K electrons, clearly only two such transitions could occur without replenishment of the K electrons. If the outer shells of the atom are still filled with electrons this causes no difficulty, since even for an atomic number as low as 10, a vacancy in the K shell will be filled by an L-shell electron in a time of the order of 10^{-14} sec., and this time will decrease rapidly (as Z^{-4}) for increase of Z. It is possible, however, to conceive of the capture of the meson in an orbit of considerably greater radius than that of the K shell. In the process of transition down to orbits near the K shell it could happen that most of the outer electrons are removed by Auger processes, so that when the K electrons have also been removed, only radiative transitions are possible. In this case, however, the atom could capture electrons from other atoms by charge exchange processes. Assuming a cross-section for such a process of the order of 10^{-16} cm.^{2}, an ionized gas atom could capture an electron in a time of about 3×10^{-9} sec., thus making further Auger transition possible.* But even if only radiative transitions were possible, Table XXIV indicates that the time t_2 of transition to the lowest state is unlikely to be greater than 10^{-10} sec. Since it is found experimentally that τ' is of the order of 10^{-6} or 10^{-7} sec., it seems established that the μ^--meson must spend the greater part of its life bound to an atom in its lowest possible $1s$ orbit. As has already been pointed out, a meson in such an orbit in an atom of medium or large atomic number will spend a considerable proportion of its time inside the nucleus. The fact that it spends such a long time in this orbit accordingly implies that the interaction between the μ-meson and the nucleus is extraordinarily weak.

Precisely similar processes will occur in the capture of π^--mesons by atomic systems. The lifetime of the π-meson, however ($2 \cdot 6 \times 10^{-8}$

---

* Alternatively, the possibility that the meson could be transferred to another atom by a kind of charge exchange process should also be considered.

TABLE XXIV. *Transition rates for μ⁻-mesons inside the K-electron orbits*

The numbers given in this table are the number of transitions per second

Quantum numbers of initial and final states $n, l \rightarrow n', l'$	Type of transition	Atomic number Z				
		2	5	10	20	40
15, 14→14, 13	Radiative	$5{\cdot}20 \times 10^{7}$	$2{\cdot}03 \times 10^{9}$	$3{\cdot}25 \times 10^{10}$	$5{\cdot}2 \times 10^{11}$	$8{\cdot}32 \times 10^{12}$
10, 9→9, 8	Radiative	$4{\cdot}08 \times 10^{8}$	$1{\cdot}60 \times 10^{10}$	$2{\cdot}56 \times 10^{11}$	$4{\cdot}1 \times 10^{12}$	$6{\cdot}55 \times 10^{13}$
9, 8→8, 7	Radiative	$7{\cdot}00 \times 10^{8}$	$2{\cdot}74 \times 10^{10}$	$4{\cdot}40 \times 10^{11}$	$7{\cdot}0 \times 10^{12}$	$1{\cdot}13 \times 10^{14}$
8, 7→7, 6	Radiative	$1{\cdot}28 \times 10^{9}$	$5{\cdot}00 \times 10^{10}$	$8{\cdot}00 \times 10^{11}$	$1{\cdot}3 \times 10^{13}$	$2{\cdot}05 \times 10^{14}$
7, 6→6, 5	Radiative	$2{\cdot}55 \times 10^{9}$	$9{\cdot}96 \times 10^{10}$	$1{\cdot}60 \times 10^{12}$	$2{\cdot}6 \times 10^{13}$	$4{\cdot}10 \times 10^{14}$
7, 6→6, 5	Auger	—	—	$3{\cdot}5 \times 10^{14}$	$5{\cdot}7 \times 10^{13}$	$8{\cdot}0 \times 10^{14}$
6, 5→5, 4	Radiative	$5{\cdot}68 \times 10^{9}$	$2{\cdot}21 \times 10^{11}$	$3{\cdot}54 \times 10^{12}$	$4{\cdot}0 \times 10^{14}$	$9{\cdot}06 \times 10^{14}$
6, 5→5, 4	Auger	—	—	3×10^{14}	$1{\cdot}5 \times 10^{14}$	$4{\cdot}1 \times 10^{14}$
5, 4→4, 3	Radiative	$1{\cdot}47 \times 10^{10}$	$5{\cdot}72 \times 10^{11}$	$9{\cdot}16 \times 10^{12}$	$1{\cdot}7 \times 10^{14}$	$1{\cdot}8 \times 10^{14}$
5, 4→4, 3	Auger	—	—	$1{\cdot}6 \times 10^{14}$	$4{\cdot}8 \times 10^{14}$	$7{\cdot}6 \times 10^{15}$
4, 3→3, 2	Radiative	$4{\cdot}76 \times 10^{10}$	$1{\cdot}86 \times 10^{12}$	3×10^{13}	$5{\cdot}4 \times 10^{13}$	$5{\cdot}5 \times 10^{13}$
4, 3→3, 2	Auger	$1{\cdot}0 \times 10^{13}$	$4{\cdot}0 \times 10^{13}$	$5{\cdot}0 \times 10^{13}$	$2{\cdot}25 \times 10^{15}$	$3{\cdot}6 \times 10^{16}$
3, 2→2, 1	Radiative	$2{\cdot}23 \times 10^{11}$	$8{\cdot}68 \times 10^{12}$	$1{\cdot}4 \times 10^{14}$	$9{\cdot}0 \times 10^{12}$	$9{\cdot}1 \times 10^{12}$
3, 2→2, 1	Auger	—	$6{\cdot}0 \times 10^{12}$	$1{\cdot}35 \times 10^{15}$	$2{\cdot}15 \times 10^{16}$	$3{\cdot}5 \times 10^{17}$
2, 1→1, 0	Radiative	$2{\cdot}16 \times 10^{12}$	$8{\cdot}44 \times 10^{12}$	$2{\cdot}8 \times 10^{11}$	$3{\cdot}1 \times 10^{11}$	$3{\cdot}2 \times 10^{11}$
2, 1→1, 0	Auger	—	$2{\cdot}1 \times 10^{11}$	—	$1{\cdot}4 \times 10^{12}$	2×10^{15}
2, 0→2, 1	Radiative	—	$2{\cdot}4 \times 10^{4}$	$2{\cdot}3 \times 10^{8}$	2×10^{8}	$5{\cdot}2 \times 10^{9}$
2, 0→1, 0	Auger*	—	$5{\cdot}2 \times 10^{9}$	$5{\cdot}2 \times 10^{9}$	$5{\cdot}2 \times 10^{9}$	5×10^{12}
2, 0→1, 0	Internal pairs	—	—	—	—	—
8, 6→6, 5	Radiative	—	$2{\cdot}9 \times 10^{10}$	$4{\cdot}68 \times 10^{11}$	$7{\cdot}5 \times 10^{12}$	$1{\cdot}2 \times 10^{14}$
8, 3→3, 2	Radiative	—	$2{\cdot}03 \times 10^{11}$	$3{\cdot}25 \times 10^{12}$	$5{\cdot}2 \times 10^{13}$	$8{\cdot}3 \times 10^{14}$
8, 1→1, 0	Radiative	—	$1{\cdot}11 \times 10^{12}$	$1{\cdot}79 \times 10^{13}$	$2{\cdot}9 \times 10^{14}$	$4{\cdot}6 \times 10^{15}$
15, 13→13, 12	Radiative	—	$3{\cdot}42 \times 10^{6}$	$5{\cdot}47 \times 10^{7}$	$8{\cdot}8 \times 10^{8}$	$1{\cdot}4 \times 10^{10}$
15, 7→7, 6	Radiative	—	$6{\cdot}37 \times 10^{8}$	$1{\cdot}02 \times 10^{10}$	$1{\cdot}6 \times 10^{11}$	$2{\cdot}6 \times 10^{12}$
15, 1→1, 0	Radiative	—	$1{\cdot}67 \times 10^{11}$	$2{\cdot}68 \times 10^{12}$	$4{\cdot}3 \times 10^{13}$	$6{\cdot}9 \times 10^{13}$

* The abnormally low values for this transition, calculated by Wheeler, arose from the existence of a node in the 2s wave function causing very considerable cancellation in the transition matrix element.

sec.), is much shorter than that of the μ-meson. A large number of 'stars' induced by π^--mesons are observed in photographic emulsions. Evidently these are produced when a meson moving in the lowest orbit is captured by the nucleus. Since the lifetime of the π^--meson against decay to a μ^--meson is so short, the fact that an appreciable number of 'meson-induced stars' are observed evidently implies that the nuclear interaction of the meson is large.

It is clear from the above discussion that a considerable number of Auger electrons should be ejected from an atom that has captured a meson (π^- or μ^-). Many of these electrons will, however, be ejected with very little energy and will thus escape detection by the photographic emulsion technique. Auger electrons of energy greater than 20 keV. should be capable of detection. The calculations of de Borde would suggest that for the light elements (C, O, N) in the photographic emulsion 12 % of the negative mesons reaching the 1s meson orbit should give an Auger electron of energy between 19 and 25 keV., while about 0·5 % should give an Auger electron of energy greater than 100 keV.

For the heavy elements (Ag, Br, I) in the emulsion about 60% of the negative μ mesons should give one Auger electron of energy greater than 20 keV., while 5 % of them should give two such electrons. About 3 % of them would be expected to give rise to an electron of energy greater than 100 keV. and about 0·1 % to one of energy about 300 keV.

Cosyns and collaborators (1949) did not find quite as many cases of electrons of energy greater than 20 keV. accompanying negative meson capture or decay as would be expected from Burbidge's calculations. They observed altogether 413 μ-mesons ending in the emulsion. Of these 152 gave rise neither to fast-decay electrons nor to Auger electrons, 230 to fast electrons only, 2 to a fast electron and an Auger electron, and 29 to an Auger electron (or electrons) only. Assuming that half the mesons were positively charged and thus gave rise to fast electrons only, and applying a correction for loss of positive decays owing to failure to observe some of the fast electrons, it appears that the 230 observed decays should be divided into 48 negative and 182 positive decays.* If the μ^--mesons cap-

* I am indebted to Dr O. Rochat for this estimate of the 'loss' correction of positive decays.

tured by heavy atoms can be distinguished from those captured by light atoms by the fact that in the former case there is no decay electron, the observation of Cosyns *et al.* show that about 29 out of 156 (i.e. 18·5 %) of the μ^--mesons captured by Ag, Br or I give Auger electrons of energy greater than 20 keV. The calculation would suggest that about 60% should have associated Auger electrons.

Again, of 54 μ^--mesons absorbed in light nuclei only 2 (i.e. 4 %) gave rise to Auger electrons while the calculations suggest that 12 % should do so. The results of Franzinetti (1950) and Fry (1950) are consistent with those of Cosyns.

The comparison appears reasonable when it is remembered that the calculations assume the mesons are captured into circular orbits of high total quantum number and make transitions to lower circular orbits.

For heavier mesons the proportion of Auger electrons is slightly higher. Thus for a meson of mass equal to 1000 electron masses captured by Ag or Br, the proportion of capture processes giving rise to Auger electrons of energy above 20 keV. should be about 50% greater than for μ meson capture.

PLATE V

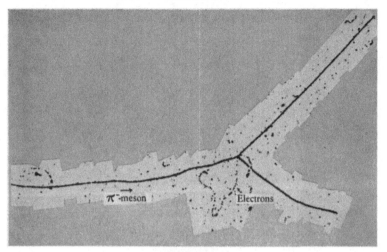

π^--meson   Electrons

$\leftarrow\ \overline{}\ \rightarrow$
60μ

(a)

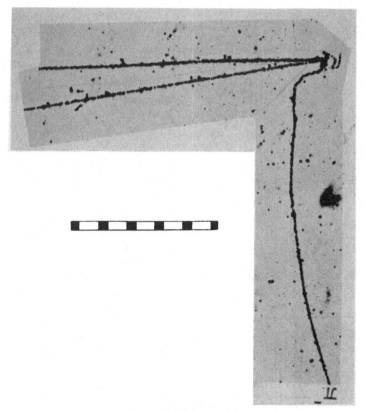

(b)

For explanation see p. xii

CHAPTER VIII

RADIATIONLESS TRANSITIONS IN ATOMIC AND MOLECULAR SPECTRA

8.1. Auto-ionization

Processes very similar to the Auger effect are of importance in the interpretation of certain width anomalies in atomic spectra. In some cases an excited discrete state of an atom may have an energy in excess of the ionization energy. A two-electron radiationless transition is then possible, in which one electron is ejected from the atom and another makes a transition to a level of smaller energy. An Auger-type process, when associated with the outer levels of the atom which determine atomic spectra, is known as auto-ionization.

Many examples of auto-ionization are provided by a study of the arc spectrum of copper. In its normal state the copper atom has the outer-electron configuration $3d^{10}\,4s$. The $4s$ electron, being the first electron in the formation of a new shell, is not very tightly bound. An excited configuration $3d^9\,4s\,5s$ exists for which the excitation energy exceeds that needed to remove the electron from the normal atom. A transition is possible in which the $3d$ shell is completed and the $4s$ electron ejected. The possibility of such a transition greatly reduces the lifetime of many of the states with the configuration $3d^9\,4s\,5s$. As a result their breadth is markedly increased, and emission lines having them as initial state are unusually wide.

The importance of auto-ionization in atomic spectra was first recognized by Shenstone (1931) on the basis of some careful measurements by Allen (1931) of the widths of a number of Cu I lines. Allen found certain lines of much greater width than could be explained by Döppler effect or pressure broadening. Fig. 48 shows the results of measurements, over a wide range of pressure, of the breadths of various lines having as initial state $3d^9\,4s\,5s\,(^4D)$. The lines $\lambda 4509$, $\lambda 4275$, for which the $^4D_{\frac{1}{2}}$ and $^4D_{3\frac{1}{2}}$ are initial levels, are seen to have very small breadths at low pressures. On the other hand, the lines $\lambda 4539$, originating in the $^4D_{1\frac{1}{2}}$ level, and $\lambda 4587$ and

$\lambda 4378$, originating in the $^4D_{2\frac{1}{2}}$ level, have substantial widths even at low pressure. From his measurements Allen was able to estimate natural widths of 0·37, 4·48 and 2·50 cm.$^{-1}$ for the $^4D_{\frac{1}{2}}$, $^4D_{1\frac{1}{2}}$ and $^4D_{2\frac{1}{2}}$ levels.

Allen's observation that certain of the levels of the multiplet show evidence of auto-ionization while others do not, indicates the existence of selection rules governing radiationless transitions of this type which may forbid the transition even when it is energeti-

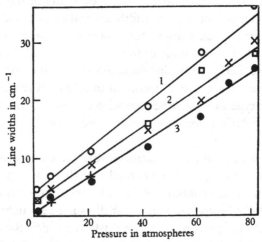

Fig. 48. Variation with pressure of the breadth of lines in the copper arc spectrum arising from transitions from the $3d^9 4s 5s (^2D)$ state. (1) $\lambda 4539$ (upper state $^4D_{1\frac{1}{2}}$); (2) $\lambda 4587$ and $\lambda 4378$ (upper state $^4D_{2\frac{1}{2}}$); (3) $\lambda 4509$ (upper state $^4D_{\frac{1}{2}}$) and $\lambda 4275$ (upper state $^4D_{3\frac{1}{2}}$).

cally possible. For example, in the case of the $3d^9 4s 5s (^4D)$ state, the Cu$^+$ ion after the transition has the configuration $3d^{10}$, so that the only possible states of the final system of Cu$^+$ together with ejected electron are doublet states. If parity and total angular momentum (J) are to be conserved the more important final states will therefore be $^2D_{1\frac{1}{2}}$ and $^2D_{2\frac{1}{2}}$.*

The selection rules can be deduced from a theory of the auto-ionization process which would follow similar lines to that of the Auger effect outlined in Chapters II and IV. A consideration of the expression (2.9) shows that the parity of the initial and final states

* Since L-S coupling evidently breaks down in this case the final states, $^2S_{\frac{1}{2}}$, $^2G_{3\frac{1}{2}}$, would also be possible but are likely to be improbable since they involve a large change in L.

of the system must be the same.* Consideration of the angular integrals such as (4.9) and (4.10) of Chapter IV leads to the selection rules that $\Delta J = 0$ for radiationless transitions, while, if spin-orbit interactions are unimportant, the selection rules $\Delta L = 0$, $\Delta S = 0$ apply also.

Although the most detailed study of line broadening due to auto-ionization has been carried out for the Cu I spectrum, examples of a similar phenomenon are known in many other spectra.†

Auto-ionization can frequently occur from states of double excitation. For example, the excitation of both electrons of helium into $2s$ levels can give rise to an auto-ionization process in which one of the electrons drops back to a $1s$ level and the other has sufficient energy for ejection from the atom. Detailed calculations have been made of the auto-ionization transition rate of states of double excitation of helium by Kiang, Ma and Wu (1936) and by Wu (1944). For the states $2s^2\,(^1S)$, $2s\,2p\,(^3P)$, $2p^3\,(^1D)$ and $3d^2\,(^1G)$ transition rates of 4×10^{14}, 5×10^{13}, 10^{14} and 4.9×10^{13} per sec., respectively, were obtained, corresponding to level widths of 0·26, 0·033, 0·065 and 0·033 eV. respectively. Emission lines originating in these doubly-excited states would thus be expected to be very broad. Two lines ($\lambda\,320\cdot4$, $\lambda\,357\cdot5$) in the far ultra-violet spectrum of helium, observed by Compton and Boyce (1928) and Kruger (1930), were attributed to transitions from doubly-excited states of helium. These lines, however, were observed to be quite sharp, so that it is difficult to reconcile them with the calculation of Wu which would predict their width to be of the order of 1 Å. However, they may arise from an initial state for which auto-ionization is forbidden by the selection rules.‡

The high transition rate for auto-ionization from doubly-excited states of helium might be thought to be of some importance in electrical discharge phenomena by providing a process by which

* The parity of an atomic system is even or odd according as $\Sigma_i\, l_i$ summed over all the electrons is even or odd, l_i being the azimuthal quantum number of the i^{μ} electron.

† See Herzberg (1944, p. 167). Most examples of broadening by auto-ionization are to be found in the Cu I spectrum however (Shenstone, 1948).

‡ Similar calculations have been made by Wu and Ourom (1950) for radiation-less transitions from the $1s^2 2p\,3s\,(^3P)$ state of Be, but the calculated width is much larger than that observed.

the ionization rate of helium could be augmented. However, the cross-section for double excitation of helium in electron collisions turns out to be very small. Thus the cross-section for excitation to the $2s\,2p\,(^1P)$ level has been calculated by Massey and Mohr (1933) to be 6×10^{-19} cm.^{2} for electrons of energy 400 eV. compared with a cross-section of about $2 \cdot 5 \times 10^{-17}$ cm.^{2} for direct ionization. Similar small values of the cross-section for double excitation by electron impact have been obtained by Wu and Yu (1944) in calculations for lithium.

It is of interest to note that an auto-ionization process was suggested by Shenstone (1931) to explain the existence of ultra-ionization potentials for mercury. Very careful measurements of the variation with electron energy of the ionization cross-section for mercury vapour by electron impact have established the existence of a number of maxima at electron energies a few volts above the ionization potential. Shenstone suggested that these might be associated with the onset of ionization through an auto-ionization process after excitation to a level having an excitation energy greater than the ionization energy of mercury. This explanation has not been put to a quantitative test owing to the difficulty of calculating ionization and excitation cross-sections for electrons of energy close to the excitation energy.*

Evidence of auto-ionization has also been obtained in molecular spectra in both absorption and emission. Absorption bands show a diffuseness in certain cases which can be ascribed to the fact that the upper level lies above the ionization limit of the molecule, and radiationless transitions are possible from it in which electrons are ejected from the molecule. At the same time emission bands from the upper state are weakened owing to the competing process of auto-ionization. Beutler and Jünger (1936) have found evidence for auto-ionization in the hydrogen molecular spectrum in both emission and absorption.

8.2. Perturbations in molecular spectra

It is well known that certain anomalies are observed in the rotational structure of bands observed under high dispersion. For

---

* For further details of ultra-ionization potentials see Massey and Burhop (1952, p. 42).

example, the frequency of the individual lines of a band is given in general by a relation of the type $\nu = A + BJ + CJ^2$, where A, B, C are constants and J takes integral values. Sometimes, however, the regular progression of the lines in the band is broken. For a short range of values of J the relation breaks down, while for both larger and smaller values it represents the positions of the lines in the band quite well. Such a phenomenon is referred to as a perturbation. It may arise from a displacement of some of the rotational energy levels of either the upper or lower electronic state concerned in the transition owing to the existence of another electronic state having rotational levels of very nearly the same energy. As a result radiationless transitions to and fro can occur between the two levels of nearly equal energy, and a molecule in one or other of these rota-

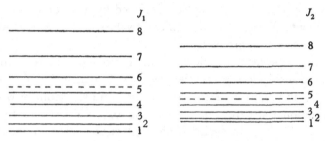

Fig. 49. Illustrating perturbations in molecular spectra.

tional states must be considered as spending part of its time in the other. Its wave function is represented by a mixture of the wave function of the two states (in the absence of the perturbation) and its energy is altered. Fig. 49 shows two systems of rotational states in which the unperturbed level $J = 5$ of (a) lies very close to, but slightly above the level $J = 5$ of (b). As a result the positions of the levels in the two cases are changed to those shown by the dotted lines. The level $J = 5$ of (a) is displaced toward higher energy, while that of (b) is displaced toward lower energies by an equal amount.

Vibrational perturbations are also observed in which the whole of a vibrational level is displaced from its normal position in a progression of bands owing to the presence of a vibrational level of another electronic state of nearly the same energy for all rotational levels.

For further discussion of this class of radiationless transition the reader is referred to works on molecular spectra such as Herzberg (1950, p. 280) or Kronig (1930).

8.3. Predissociation

Another radiationless transition phenomenon important in the interpretation of molecular spectra is that of predissociation. It has similar consequences to auto-ionization, and arises when certain excited vibrational and rotational levels of a given electronic state have a greater energy than the dissociation limit of the molecule in another electronic state. Radiationless transition can then occur from the former levels, leading to dissociation of the molecule. The phenomenon could also be regarded as a perturbation in which the discrete levels of (b) (fig. 49) are replaced by a continuum corresponding to dissociation with different values of the kinetic energy of the dissociation products.

As in auto-ionization a state leading to predissociation is broadened. In absorption spectra the presence of predissociation reveals itself by an abnormal diffuseness of the band spectrum in which the rotational structure of the band is lost. This is because the lifetime of the state against radiationless transition to the dissociated state is usually short compared with the time of molecular rotation. The vibrational structure is generally preserved, however, since the vibrational period of the molecule is usually much shorter than the time of molecular rotation. The diffuseness in the absorption band structure may set in quite sharply at a certain point of a band progression. Thus, proceeding toward higher frequencies in a band progression, the first bands may have sharp lines, but at a certain stage the rotational structure may become diffuse. This arises from the fact that for a given final vibrational state of the molecule, the vibrational quantum number of the initial state increases in the direction of increasing frequency of the band. It may happen that although the lowest vibrational levels may correspond to a total energy below the dissociation limit of an electronic state, levels of higher vibrational quantum number have a total energy in excess of this limit. Predissociation is energetically possible only for these latter levels, so that the characteristic band diffuseness sets in quite suddenly at a certain stage in the progression.

In some cases the rotational structure may be quite sharp for the early lines in a given absorption band but may become diffuse for lines corresponding to transitions from levels of larger rotational quantum number J. A structure of this type, partly sharp and partly diffuse, may persist for several successive bands in a progression, the demarcation between sharp and diffuse structure coming closer to the band head in successive bands. Evidently the states of small J are not sufficiently energetic to exhibit predissociation, while those of the larger J can do so.

In emission spectra predissociation reveals itself by a sudden weakening or, more frequently, disappearance of bands in a given progression for a certain value of the vibrational quantum number of the initial state. Similarly, abrupt termination of a single band may occur at a certain value of J, above which predissociation is possible. The disappearance of the rest of the band in a case of this kind evidently indicates that the probability of radiationless transition leading to dissociation greatly exceeds the probability of radiative transitions to the lower state.

Direct evidence that dissociation does indeed accompany the anomalies attributed to predissociation in molecular spectra has been obtained by Bonhoeffer and Farkas (1927). They illuminated NH_3 with light of the correct frequency to bring the molecule (by absorption) into the upper state of some bands exhibiting the phenomenon of predissociation and detected the formation of H_2 and N_2.

Just as for auto-ionization, predissociation cannot always occur when it is energetically possible. Certain selection rules have to be satisfied relating the initial state of predissociation and the final state in which dissociation occurs. They must have the same parity and the same total angular momentum, and, for the case of homonuclear diatomic molecules, the same symmetry. Other selection rules may also apply depending on the type of coupling.

In addition, the Franck-Condon principle continues to hold for the radiationless transition leading to dissociation. Thus transitions can only occur provided the nuclear separation in the final state is almost the same as in the initial state. This means that the potential energy curves of the two states in question must nearly intersect, and transitions only occur from vibrational levels of the initial state which have an energy close to that of the point of near-intersection.

A considerable body of experimental information concerning predissociation has now been accumulated, but its detailed discussion falls outside the scope of the present work. The reader is referred to works on molecular spectra such as Herzberg (1950, p. 405) for further information.

Radiationless transition processes in molecules are of interest in fields far wider than that of the interpretation of molecular spectra. For example, Bates (1950) has recently discussed the probability of the dissociative recombination process $XY^+ + e \rightarrow X' + Y'$, in which recombination takes place between an ionized molecule XY^+ and an electron, as a result of which a neutral molecule XY is formed in a state of predissociation leading to dissociation into the excited atoms X' and Y'. He concluded that such a process could, under some circumstances, lead to a very high value of the electron-ion recombination coefficient, and might account for abnormally high recombination coefficients deduced in recent measurements of recombination rates in rare gases (Biondi and Brown, 1949; Holt, Richardson, Howland and McClure, 1950). Similar processes are believed to be of considerable importance in the formation of negative atomic ions by dissociative attachment of electrons in a molecular gas.*

8.4. Other radiationless transition processes

The topics treated in this book by no means exhaust the range and variety of physical phenomena in which radiationless transitions play an important role. No reference has been made to the very important class of phenomena which come under the head of collisions of the second kind, nor to the interaction of positive ions and excited atoms with metal surfaces, nor to examples of radiationless transitions arising in the theory of solids. As pointed out in Chapter 1 many of these topics are treated in some detail elsewhere.

The main purpose of the book has been, however, to discuss radiationless transition processes associated with the inner atomic levels and the nucleus where these processes play such an important role.

* These processes are discussed by Massey (1950, p. 58).

REFERENCES AND AUTHOR INDEX

AGENO (1943). *Nuovo Cim.* **1**, 415.   *pages* 130
ALICHANOW (1938). *Bull. Acad. Sci. U.R.S.S.* **1-2**, 33.   155
ALICHANOW, ALICHANIAN and KOSODAEW (1936). *J. Phys. Radium,* **7**, 163.   146, 152
ALICHANOW and DZELEPOW (1938). *C.R. Acad. Sci. U.R.S.S.* **20**, 113.   146, 152, 154
ALICHANOW and LATYSHEV (1938). *C.R. Acad. Sci. U.R.S.S.* **20**, 429.   146, 152
ALLEN (1931). *Phys. Rev.* **39**, 42.   171
ALVAREZ (1938). *Phys. Rev.* **54**, 486.   129
ARENDS (1935). *Ann. Phys., Lpz.,* **22**, 281.   37, 45, 48
ARNOULT (1939). *Ann. Phys., Paris,* **12**, 240.   23, 62, 136
AUGER (1925). *J. Phys. Radium,* **6**, 205.   2, 24, 45, 55
AUGER (1926). *Ann. Phys., Paris,* **6**, 183.   2, 24, 26, 48, 52
AXEL and DANCOFF (1949). *Phys. Rev.* **76**, 892.   140, 141
BACKHURST (1936). *Phil. Mag.* **22**, 737.   45, 47, 48
BALDERSTONE (1926). *Phys. Rev.* **27**, 696.   45, 47
BARKLA and PHILPOT (1913). *Phil. Mag.* **25**, 849.   3, 37, 45, 47
BARKLA and SADLER (1917). *Philos. Trans.* A, **217**, 315.   3
BARTLETT (1930). *Phys. Rev.* **35**, 229.   80
BATES (1950). *Phys. Rev.* **78**, 492.   178
BEARDEN and SNYDER (1941). *Phys. Rev.* **59**, 162.   53, 91
BEATTY (1911). *Proc. Roy. Soc.* A, **85**, 329.   3, 37, 45, 47
BENNETT (1938). *Proc. Camb. Phil. Soc.* **34**, 282.   137
BERESTETZKY (1947). *J. Phys. U.S.S.R.* **11**, 85.   100, 118
BERESTETZKY and SHMUSHKEVICH (1950). *J. Exp. Theor. Phys. U.S.S.R.* **20**, 574.   150
BERGSTRÖM (1950). *Phys. Rev.* **80**, 114.   132
BERGSTRÖM and THULIN (1949). *Phys. Rev.* **76**, 1718.   129
BERGSTRÖM and THULIN (1950a). *Phys. Rev.* **79**, 537.   129
BERGSTRÖM and THULIN (1950b). *Phys. Rev.* **79**, 538.   133
BERGSTRÖM, THULIN and ANDERSSON (1950). *Phys. Rev.* **77**, 851.   129
BERKEY (1934). *Phys. Rev.* **45**, 437.   45, 47
BERTHELOT (1944). *Ann. Phys., Paris,* **19**, 117, 219.   103, 129, 140
BETHE (1937). *Rev. Mod. Phys.* **2**, 223.   103
BEUTLER and JUNGER (1936). *Z. Phys.* **100**, 80; **101**, 285.   174
BIONDI and BROWN (1949). *Phys. Rev.* **75**, 1700; **76**, 1697.   178
BLOCH, F. (1935). *Phys. Rev.* **48**, 187.   90
BLOCH and ROSS (1935). *Phys. Rev.* **47**, 884.   90
BOEHM, BLASER, MARMIER and PREISWERK (1950). *Phys. Rev.* **77**, 295.   128
BOEHM, HUBER, MARMIER, PREISWERK and STEFFEN (1949). *Helv. phys. Acta,* **22**, 69.   131
BOEHM and PREISWERK (1949). *Helv. phys. Acta,* **22**, 331.   131

BONHOEFFER and FARKAS (1927). *Z. phys. Chem.* A, **134**, 337.   *pages* 177
BOTHE (1925). *Phys. Z.* **26**, 410.   4
BOWE, GOLDHABER, HILL, MEYERHOF and SALA (1948). *Phys. Rev.*
   73, 1219.   142
BOWE and SCHARFF-GOLDHABER (1949). *Phys. Rev.* **76**, 437.   132
BOWER (1936). *Proc. Roy. Soc.* A, **157**, 662.   24, 52, 55
BRADT, GUGELOT, HUBER, MEDICUS, PREISWERK, SCHERRER and
   STEFFEN (1946). *Helv. phys. Acta*, **19**, 218.   130, 131
BRADT, HEINE and SCHERRER (1943). *Helv. phys. Acta*, **16**, 455.   137
BRADT, HALTER, HEINE and SCHERRER (1946). *Helv. phys. Acta*,
   19, 431.   146, 155
BRADT and SCHERRER (1945). *Helv. phys. Acta*, **18**, 260, 405.   137
BRADT and SCHERRER (1946). *Helv. phys. Acta*, **19**, 307.   137
BRADY and DEUTSCH (1947). *Phys. Rev.* **72**, 870.   127
BRADY and DEUTSCH (1950). *Phys. Rev.* **78**, 558.   125, 156
BRIL (1947). *Physica*, **13**, 481.   82, 86
BUNKER and CANADA (1950). *Phys. Rev.* **80**, 961.   129
BURBANK (1939). *Phys. Rev.* **56**, 142.   89
BURBIDGE (1951). Thesis, London.   166
BURHOP (1934). *Proc. Roy. Soc.* A, **145**, 612.   90
BURHOP (1935). *Proc. Roy. Soc.* A, **148**, 272.   47, 49
BURHOP (1940). *Proc. Camb. Phil. Soc.* **36**, 43.   76, 91
CADY and TOMBOULIAN (1941). *Phys. Rev.* **59**, 381.   95
CALDWELL (1950). *Phys. Rev.* **78**, 407.   128, 129, 133, 134
CAUCHOIS (1943). *C.R. Acad. Sci., Paris*, **216**, 529.   91
CAUCHOIS (1944). *J. Phys. Radium*, **5**, 1.   75
CAUCHOIS and HULUBEI (1947). *Longuers d'onde des émissions X
   et des discontinuités d'absorption X*. Paris: Hermann.   75
CHANG (1949). *Rev. Mod. Phys.* **21**, 166.   162
CHU and WIEDENBECK (1949). *Phys. Rev.* **75**, 226.   121, 133
COISH (1951). *Phys. Rev.* **84**, 164.   20
COMPTON (1929). *Phil. Mag.* **8**, 961.   45, 47, 48
COMPTON and BOYCE (1928). *J. Franklin Inst.* **205**, 497.   173
CONSTANTINOV and LATYSHEV (1941). *J. Phys. U.S.S.R.* **5**, 249.   137
COOPER (1942). *Phys. Rev.* **61**, 234.   53, 82, 86
COOPER (1944). *Phys. Rev.* **65**, 155.   74, 87
CORK and SMITH (1941). *Phys. Rev.* **60**, 480.   133
COSTER and BRIL (1942). *Physica*, **9**, 84.   82, 86
COSTER and KRONIG (1935). *Physica*, **2**, 13.   4, 68
COSTER, KUIPERS and HUIZINGA (1935). *Physica*, **2**, 870.   79
COSTER and DE LANGEN (1936). *Physica*, **3**, 282.   82
COSTER and DE LANGEN (1947). *Physica*, **13**, 385.   53
COSYNS, DILWORTH and OCCHIALINI, SCHOENBERG and PAGE
   (1949). *Proc. Phys. Soc.* A, **62**, 801.   162, 169
CRANBERG (1950). *Phys. Rev.* **77**, 155.   136
CREUTZ, DELSASSO, SUTTON, WHITE and BARKAS (1940). *Phys.
   Rev.* **58**, 481.   129, 132
CURRAN, ANGUS and COCKCROFT, A. L. (1949).
   Phil. Mag. **40**, 36.   24, 28, 30, 45, 83, 115
DANCOFF and MORRISON (1939). *Phys. Rev.* **55**, 122.   117, 118, 122

DE BROGLIE and THIBAUD (1925). *C.R. Acad. Sci., Paris,* **180**, 179. *pages* 57
DE LANGEN (1939). *Physica,* **6**, 27. 82
DE LANGEN (1940). *Physica,* **7**, 845. 82
DEUTSCH, ELLIOTT and EVANS (1944). *Rev. Sci. Instrum.* **15**, 178. 109
DEUTSCH, ELLIOTT and ROBERTS (1945). *Phys. Rev.* **68**, 193. 128
DEUTSCH and HEDGRAN (1949). *Phys. Rev.* **75**, 1443. 128
DEUTSCH and SIEGBAHN, K. (1950). *Phys. Rev.* **77**, 680. 128
DE VAULT and LIBBY (1940). *Phys. Rev.* **58**, 688. 115
DEVONS (1949). *Excited States of Nuclei,* p. 92. Cambridge University Press. 141
DEVONS, HEREWARD and LINDSAY (1949). *Nature, Lond.,* **164**, 586. 158
DEVONS and LINDSAY (1949). *Nature, Lond.,* **164**, 539. 152, 157, 158
DRELL (1949). *Phys. Rev.* **75**, 132. 118, 123
DUNWORTH (1940). *Rev. Sci. Instrum.* **11**, 167. 127
ELLIS (1933a). *Proc. Roy. Soc.* A, **139**, 336. 48, 51, 61
ELLIS (1933b). *Proc. Roy. Soc.* A, **143**, 350. 137
ELLIS and ASTON (1930). *Proc. Roy. Soc.* A, **129**,
180. 113, 121, 136, 137, 141
ELLIS and SKINNER (1924). *Proc. Roy. Soc.* A, **105**, 185. 8
FEATHER (1940). *Proc. Camb. Phil. Soc.* **36**, 224. 107
FEATHER and DUNWORTH (1938). *Proc. Roy. Soc.* A, **168**, 566. 130
FEATHER, KYLES and PRINGLE (1948). *Proc. Phys. Soc.*
61, 466. 107, 108, 115
FERENCE (1937). *Phys. Rev.* **51**, 720. 59, 63
FERMI and TELLER (1947). *Phys. Rev.* **72**, 399. 164
FIERZ (1945). *Helv. phys. Acta,* **16**, 365. 103
FISK (1934). *Proc. Roy. Soc.* A, **143**, 674. 117
FISK and TAYLOR (1934). *Proc. Roy. Soc.* A, **146**, 178. 102, 117, 121, 122
FLAMMERSFELD (1939). *Z. Phys.* **114**, 227. 7, 23, 48, 51, 62, 113, 115, 136
FLÜGGE (1941). *Ann. Phys., Lpz.,* **39**, 373. 102, 103
FOWLER (1930). *Proc. Roy. Soc.* A, **129**, 1. 21, 141
FRANZINETTI (1950). *Phil. Mag.* **41**, 86. 162, 170
FRASER (1949). *Phys. Rev.* **76**, 1540. 133
FRAUENFELDER, GUGELOT, HUBER, MEDICUS, PREISWERK, SCHERRER
and STEFFEN (1948). *Phys. Rev.* **73**, 1270. 134
FRAUENFELDER, HUBER, DE SHALIT and ZÜNTI (1950). *Phys. Rev.* **79**,
1029. 135
FRAUENFELDER, WALTER and ZÜNTI (1950). *Phys. Rev.* **77**, 557. 127
FREEDMAN and ENGELKEMEIR (1950). *Phys. Rev.* **79**, 897. 133
FREEDMAN, JAFFE and WAGNER (1950). *Phys. Rev.* **79**, 410. 135
FRIEDLANDER, PERLMAN and SCHARFF-GOLDHABER (1950). *Phys.
Rev.* **80**, 1103. 132
FRILLEY and TSIEN (1945). *C.R. Acad. Sci., Paris,* **220**, 144. 83
FRY (1950). *Phys. Rev.* **79**, 893. 162, 170
FULLER (1950). *Proc. Phys. Soc.* A, **63**, 1348. 133
GARDNER (1949). *Proc. Phys. Soc.* **62**, 763. 127, 138
GEI, LATYSHEV and PASECHNICK (1948). *Bull. Acad. Sci. U.R.S.S.,*
Physics Series 12, sect. 6. 158
GELLMAN, GRIFFITH and STANLEY (1950). *Phys. Rev.*
80, 866. 117, 120, 121, 122

GERMAIN (1950). *Phys. Rev.* **80**, 937.   *pages* 24, 30, 48, 51

GOERTZEL and LOWEN (1945). *Phys. Rev.* **67**, 203.   118

GOLDBERGER (1948). *Phys. Rev.* **73**, 1119.   143

GOLDHABER, MUEHLHAUSE and TURKEL (1947). *Phys. Rev.* **71**, 372.   143

GOLDHABER and STURM (1946). *Phys. Rev.* **70**, 111.   129

GRANT and RICHMOND (1949). *Proc. Phys. Soc.* A, **62**, 573.   133

GRINBERG and ROUSSINOW (1940). *Phys. Rev.* **58**, 181.   143

HAAS (1932). *Ann. Phys., Lpz.*, **16**, 473.   45, 46, 48

HAMILTON (1940). *Phys. Rev.* **58**, 122.   127

HANNA, KIRKWOOD and PONTECORVO (1949). *Phys. Rev.* **75**, 985.   28

HARMS (1927). *Ann. Phys., Lpz.*, **82**, 87.   45, 47, 48

HASLAM, KATZ, MOODY and SKARSGARD (1950). *Phys. Rev.* **80**, 318.   128

HAYNES (1948). *Phys. Rev.* **74**, 423.   142

HAYWARD (1950*a*). *Phys. Rev.* **79**, 541.   129

HAYWARD (1950*b*). *Phys. Rev.* **79**, 542.   131

HEBB and NELSON (1940). *Phys. Rev.* **58**, 486.   117, 123

HEBB and UHLENBECK (1938). *Physica*, **5**, 605.   103, 117, 140

HEITLER (1936). *Proc. Camb. Phil. Soc.* **32**, 112.   100

HELMHOLZ (1941). *Phys. Rev.* **60**, 415.   129, 130

HERZBERG (1944). *Atomic Spectra and Atomic Structure.* New York:
    Dover Publications.   173

HERZBERG (1950). *Molecular Spectra and Molecular Structure*, **1**.
    New York: Van Nostrand.   176, 178

HILL, J. M. and SHEPHERD (1950). *Proc. Phys. Soc.* A, **63**, 126.   133

HILL, R. D. (1949). *Phys. Rev.* **76**, 333.   132

HILL, R. D. (1950). *Phys. Rev.* **79**, 413.   134

HILL, R. D. and MIHELICH (1948). *Phys. Rev.* **74**, 1874.   131

HIRSCH (1935). *Phys. Rev.* **48**, 722.   77, 78

HIRSCH (1936). *Phys. Rev.* **50**, 191.   77

HIRSCH (1942). *Phys. Rev.* **62**, 137.   79

HIRSCH and RICHTMYER, F. K. (1933). *Phys. Rev.* **44**, 955.   78

HOLE (1948*a*). *Ark. Mat. Astr. Fys.* A, **36**, no. 2.   133, 141

HOLE (1948*b*). *Ark. Mat. Astr. Fys.* A, **36**, no. 9.   131, 133, 134, 141

HOLT, RICHARDSON, HOWLAND and McCLURE (1950). *Phys. Rev.*
    77, 239.   178

HORTON (1948). *Proc. Phys. Soc.* **60**, 457.   152

HUDGENS and LYON (1949). *Phys. Rev.* **75**, 206.   129, 130

HULME (1932). *Proc. Roy. Soc.* A, **138**, 643.   117, 121

HULME, MOTT, OPPENHEIMER, F. and TAYLOR (1936).
    Proc. Roy. Soc. A, **155**, 315.   102, 117, 121, 124

HULUBEI (1947). *C.R. Acad. Sci., Paris*, **224**, 773.   90

HULUBEI, CAUCHOIS and MANESCU (1948). *C.R. Acad. Sci., Paris*,
    226, 764.   90

ITOH and WATASE (1941). *Proc. Phys.-Math. Soc. Japan*,
    23, 142.   115, 136, 137

JAEGER and HULME (1935). *Proc. Roy. Soc.* A, **148**, 708.   146, 147, 149

JENSEN (1949). *Phys. Rev.* **76**, 958.   133

JOHANSSON (1950). *Phys. Rev.* **79**, 896.   131

KARLSSON and SIEGBAHN, M. (1934). *Z. Phys.* **88**, 76.   96

KATZ, HILL, R. D. and GOLDHABER (1950). *Phys. Rev.* **78**, 9.   131, 132

KERN, MITCHELL and ZAFFARANO (1949a). *Phys. Rev.* **75**, 1287. *pages* 131
KERN, MITCHELL and ZAFFARANO (1949b). *Phys. Rev.* **76**, 95. 132
KETELLE and PEACOCK (1948). *Bull. Amer. Phys. Soc.* no. 2, p. 42. 133
KIANG, MA and WU (1936). *Phys. Rev.* **50**, 673. 5, 173
KIESSIG (1938). *Z. Phys.* **109**, 671. 87
KINSEY (1948a). *Canad. J. Res.* A, **26**, 404. 23, 24, 46, 53, 54, 55
KINSEY (1948b). *Canad. J. Res.* A, **26**, 421. 23, 24, 39, 40, 46, 53
KNIGHT and MACKLIN (1949). *Phys. Rev.* **75**, 34. 137
KOYENUMA (1941). *Z. Phys.* **117**, 358. 103
KRONIG (1930). *Band Spectra and Molecular Structure.* Cambridge
University Press. 176
KRUGER (1930). *Phys. Rev.* **36**, 855. 173
KULCHITSKY and LATYSHEV (1941). *J. Phys. U.S.S.R.* **4**, 515. 137
KUNDU, HULT and POOL (1950). *Phys. Rev.* **77**, 71. 129
KÜSTNER and ARENDS (1935). *Ann Phys., Lpz.,* **22**,
443. 34, 37, 52, 54, 55
LANDSBERG (1949). *Proc. Phys. Soc.* A, **62**, 806. 95
LANGER, MOFFATT and PRICE (1950). *Phys. Rev.* **79**, 808. 131
LATYSHEV, GEI, BASHILOV and BARCHUK (1948). *C.R. Acad. Sci.*
U.R.S.S. **63**, 511. 158
LATYSHEV (1947). *Rev. Mod. Phys.* **19**, 132. 115, 154
LAWSON and CORK (1940). *Phys. Rev.* **57**, 982. 131
LAY (1934). *Z. Phys.* **91**, 533. 37, 45, 48, 52, 55, 57
LI (1937). *Proc. Roy. Soc.* A, **158**, 571. 136
LOCHER (1932). *Phys. Rev.* **40**, 484. 24, 45
LOWEN and TRALLI (1949a). *Phys. Rev.* **75**, 529. 118, 123
LOWEN and TRALLI (1949b). *Phys. Rev.* **76**, 1541. 118, 123
LUNDBY (1949). *Phys. Rev.* **76**, 1809. 133
McCREARY (1942). Thesis, University of Rochester (U.S.A.)
(quoted by Reitz (1950)). 121, 134
MACINTYRE (1950). *Phys. Rev.* **80**, 1018. 141
MALLARY and POOL (1950). *Phys. Rev.* **77**, 75. 131
MANN and AXEL (1950). *Phys. Rev.* **80**, 759. 129
MARTIN, D. G. E. and RICHARDSON, H. O. W. (1948). *Proc. Roy.
Soc.* A, **195**, 287. 115
MARTIN, D. G. E. and RICHARDSON, H. O. W.
(1950). *Proc. Phys. Soc.* A, **63**, 223. 113, 114, 115, 136, 137
MARTIN, D. G. E., RICHARDSON, H. O. W. and HSÜ (1948). *Proc.
Phys. Soc.* **60**, 466. 107
MARTIN, L. H. (1927). *Proc. Roy. Soc.* A, **115**, 420. 24, 37, 45, 47, 48
MARTIN, L. H., BOWER and LABY (1935). *Proc. Roy.
Soc.* A, **148**, 40. 24, 25, 44, 45, 48
MARTIN, L. H. and EGGLESTON (1937). *Proc. Roy.
Soc.* A, **158**, 46. 24, 26, 45, 48
MASSEY (1950). *Negative Ions.* Cambridge University Press. 178
MASSEY and BURHOP (1936a). *Proc. Roy. Soc.* A, **153**, 661. 47, 50
MASSEY and BURHOP (1936b). *Proc. Camb. Phil. Soc.* **32**, 461. 53, 91
MASSEY and BURHOP (1938). *Proc. Roy. Soc.* A, **167**, 53. 161
MASSEY and BURHOP (1952). *Electronic and Ionic Impact
Phenomena.* Oxford. 8, 174

MASSEY and MOHR (1933). *Proc. Roy. Soc.* A, **140**, 613.   *pages* 174

MAYO and ROBINSON (1939). *Proc. Roy. Soc.* A, **173**, 192.   57

MEDICUS, MAEDER and SCHNEIDER (1949). *Helv. phys. Acta*, **22**, 603.   129

MEDICUS, PREISWERK and SCHERRER (1950). *Helv. phys. Acta*, **23**,
299.   130

MEITNER (1922). *Z. Phys.* **9**, 131.   8

MENON, MUIRHEAD and ROCHAT (1950). *Phil. Mag.* **41**, 592.   162

METZGER (1950). *Phys. Rev.* **79**, 398.   130

METZGER and DEUTSCH (1950). *Phys. Rev.* **78**, 551.   125

MIHELICH and HILL, R. D. (1950). *Phys. Rev.* **79**, 781.   131

MIMS, HALBAN and WILSON (1950). *Nature, Lond.*, **166**, 1027.   146, 155

MITCHELL (1948). *Rev. Mod. Phys.* **20**, 296.   106

MØLLER (1931). *Z. Phys.* **70**, 786.   14

MOON, M. L., WAGGONER and ROBERTS (1950). *Phys. Rev.* **79**, 905.   128

MOTT and MASSEY (1949). *Theory of Atomic Collisions*, 2nd ed.
Oxford.   11, 13

MOTT and SNEDDON (1948). *Wave Mechanics and its Applications*.
Oxford.   69

O'BRYAN and SKINNER (1934). *Phys. Rev.* **45**, 370.   68

O'KELLEY, BARTON, CRANE and PERLMAN (1950). *Phys. Rev.* **80**, 293.   135

OPPENHEIMER, J. R. (1941). *Phys. Rev.* A, **60**, 164.   157, 158

OSABA (1949). *Phys. Rev.* **76**, 349.   133

OSBORNE and DEUTSCH (1947). *Phys. Rev.* **71**, 467.   128

OWEN, COOK and OWEN (1950). *Phys. Rev.* **78**, 686.   128

PARRATT (1938). *Phys. Rev.* **54**, 99.   86

PEACOCK, BROSI and BOGARD (1947). *Plutonium project report*,
Mon. N432, 58.   133

PEACOCK, JONES and OVERMAN (1947). *Plutonium project report*,
Mon. N432, 56.   133

PEACOCK and WILKINSON (1948). *Phys. Rev.* **74**, 297.   128

PEARSALL (1934). *Phys. Rev.* **46**, 694.   77

PINCHERLE (1934). *Mem. Accad. Lincei*, **20**, 29.   87

PINCHERLE (1935 a). *Nuovo Cim.* **12**, 81.   47, 49, 57, 63, 87

PINCHERLE (1935 b). *Nuovo Cim.* **12**, 122.   87

PINCHERLE (1935 c). *Physica*, **2**, 596.   87, 91

PINCHERLE (1942). *Phys. Rev.* **61**, 225.   81

RAE (1949). *Phil. Mag.* **40**, 1155.   146, 155

RAE (1950). *Phil. Mag.* **41**, 525.   109

RAMBERG and RICHTMYER, F. K. (1937). *Phys. Rev.* **51**, 913.   87

RANDALL and PARRATT (1940). *Phys. Rev.* **57**, 786.   78

RASMUSSEN, HORNYAK, LAURITSEN, C. C. and LAURITSEN, T.
(1950). *Phys. Rev.* **77**, 617.   156, 157

REITZ (1950). *Phys. Rev.* **77**, 10.   117, 120, 121, 134

RICHARDSON, H. O. W. (1950). *Proc. Phys. Soc.* A, **63**, 234.   113

RICHTMYER, F. K. (1937). *Rev. Mod. Phys.* **9**, 391.   78

RICHTMYER, F. K., BARNES and RAMBERG (1934).
Phys. Rev. **46**, 843.   53, 81, 84, 85, 86, 87, 91

RICHTMYER, F. K. and RAMBERG (1937). *Phys. Rev.* **51**, 925.   80

RICHTMYER, R. D. (1936). *Phys. Rev.* **49**, 1.   66

RICHTMYER, R. D. (1939). *Phys. Rev.* **56**, 146.   89

ROBERTS, ELLIOTT, DOWNING, PEACOCK and DEUTSCH (1943). *Phys. Rev.* **64**, 268. pages 132
ROBINSON (1923). *Proc. Roy. Soc.* A, **104**, 455. 2, 16, 57
ROBINSON and CASSIE (1928). *Proc. Roy. Soc.* A, **113**, 282. 2, 16, 57
ROBINSON and YOUNG (1930). *Proc. Roy. Soc.* A, **128**, 92. 2, 16, 57, 62, 63
ROSE (1949). *Phys. Rev.* **76**, 678. 147, 150, 153
ROSE, GOERTZEL, SPINRAD, HARR and STRONG (1949). *Phys. Rev.* **76**, 1883. 117, 119, 121, 125
ROSE, GOERTZEL, SPINRAD, HARR and STRONG (1951). *Phys. Rev.* **83**, 79. 117
ROSE and UHLENBECK (1935). *Phys. Rev.* **48**, 211. 148
ROSS (1926). *Phys. Rev.* **28**, 425. 31
RUTHERFORD, CHADWICK and ELLIS (1930). *Radiations from Radioactive Substances*, p. 362. Cambridge University Press. 137
SACHS (1940). *Phys. Rev.* **57**, 194. 142
SALGUEIRO and VALADARES (1949). *Portugaliae Physica*, **3**, 21. 83
SAXON (1948). *Phys. Rev.* **74**, 849. 121, 135
SAXON (unpublished, quoted by Reitz (1950). 134
SAXON and RICHARDS (1949). *Phys. Rev.* **76**, 187. 133
SCHERB and MANDEVILLE (1948). *Phys. Rev.* **73**, 1401. 134
SCHRADER (1936). Thesis, Cornell University. 86
SEGRE and HELMHOLZ (1949). *Rev. Mod. Phys.* **21**, 271. 115, 122, 141
SHENSTONE (1931). *Phys. Rev.* **38**, 873. 171, 174
SHENSTONE (1948). *Philos. Trans.* A, **241**, 207. 173
SIDAY (1941). *Proc. Roy. Soc.* A, **178**, 189. 129
SIEGBAHN, K. (1950). *Phys. Rev.* **77**, 233. 131
SIEGBAHN, M. (1931). *Spektroskopie der Röntgenstrahlen.* Berlin: Springer. 74
SKINNER (1940). *Philos. Trans.* A, **239**, 95. 92, 94, 96
SLÄTIS, DU TOIT and SIEGBAHN, K. (1950). *Phys. Rev.* **78**, 498. 131
SLIV (1949). *C.R. Acad. Sci. U.R.S.S.* **64**, 321. 158
SMITH, G. P. (1942). *Phys. Rev.* **61**, 578. 128
SPRING (1950). *Photons and Electrons.* London: Methuen. 144
STAHEL and JOHNER (1934). *J. Phys. Radium*, **5**, 97. 121, 137
STANLEY (1949). *Canad. J. Res.* A, **27**, 17. 117
STEFFEN, HUBER and HUMBEL (1949). *Helv. phys. Acta*, **22**, 167. 7, 23, 41, 46, 48, 51, 62, 63, 113, 121, 134
STEPHENSON (1937). *Phys. Rev.* **51**, 637. 31, 44, 45, 48, 52, 54, 55
STOCKMEYER (1932). *Ann. Phys., Lpz.*, **12**, 71. 37, 45, 48
STODDARD (1934). *Phys. Rev.* **46**, 837. 91
STODDARD (1935). *Phys. Rev.* **48**, 43. 91
STRATTON (1941). *Electromagnetic Theory.* New York: McGraw Hill and Co. 99
STRAUCH (1950). *Phys. Rev.* **79**, 487. 128
SUNYAR, ALBURGER, FRIEDLANDER, GOLDHABER and SCHARFF-GOLDHABER (1950). *Phys. Rev.* **79**, 181. 135
SZILARD and CHALMERS (1934). *Nature, Lond.*, **134**, 462. 115
TAYLOR and MOTT (1932). *Proc. Roy. Soc.* A, **138**, 665. 117
TAYLOR and MOTT (1933). *Proc. Roy. Soc.* A, **142**, 215. 18

THIBAUD (1925). Thèse, Paris. *pages* 141
THOMAS (1940). *Phys. Rev.* **58**, 714. 158
TOMBOULIAN (1948). *Phys. Rev.* **74**, 1887. 69
TRALLI and GOERTZEL (1951). *Phys. Rev.* **83**, 399. 20
VALADARES (1940). *Ric. Sci.* **18**, 270. 78, 81
VALADARES and MENDES (1948). *C.R. Acad. Sci., Paris*, **226**, 1185. 79
VALLEY and McCREARY (1939). *Phys. Rev.* **56**, 863. 129
VIETH and KIRKPATRICK (1939). *Phys. Rev.* **56**, 705. 89
WAGGONER (1950). *Phys. Rev.* **80**, 489. 133
WAGGONER, MOON, M. L. and ROBERTS (1950).
    Phys. Rev. **80**, 420. 112, 125, 128, 129, 133
WALKE, THOMPSON and HOLT (1940). *Phys. Rev.* **57**, 171. 128
WALTER, HUBER and ZÜNTI (1950). *Helv. phys. Acta*, **23**, 697. 127
WANG (1948). *Nature, Lond.*, **162**, 264. 147
WEISSKOPF and WIGNER (1930). *Z. Phys.* **63**, 54. 64
WEISZÄCKER (1936). *Naturwissenschaften*, **24**, 813. 102, 116
WENTZEL (1927). *Z. Phys.* **43**, 524. 11
WEST and ROTHWELL (1950). *Phil. Mag.* **41**, 873. 30, 45, 115, 129
WHEELER (1949). *Rev. Mod. Phys.* **21**, 133. 164, 165, 166
WIEDENBECK (1944). *Phys. Rev.* **66**, 36. 141
WIEDENBECK and CHU (1947). *Phys. Rev.* **72**, 1171. 133, 134
WIGHTMAN (1950). *Phys. Rev.* **77**, 521. 163, 166
WILLIAMS (1931). *Phys. Rev.* **37**, 1431. 86
WILLIAMS (1934). *Phys. Rev.* **45**, 71. 53
WU (1944). *Phys. Rev.* **66**, 291. 5, 173
WU and OUROM (1950). *Phys. Rev.* **80**, 129. 173
WU and YU (1944). *Chinese J. Phys.* **5**, 162. 174

INDEX

Printed in the United States
By Bookmasters